NATURE'S DUE

NATURE'S DUE

Healing Our Fragmented Culture

BRIAN GOODWIN

First published in 2007 by Floris Books
© 2007 Brian Goodwin
Third printing 2018

Brian Goodwin has asserted his right under the
Copyright, Designs and Patents Act 1988
to be identified as the Author of this Work

British Library CIP Data available
ISBN 978-086315-596-3
Printed by Lightning Source

I dedicate this book to the Christel River
that changed the course of my life

Contents

Acknowledgments

Nature's Due emerged from the learning process I went through while teaching and doing research at Schumacher College during the past five years. I am grateful to all those who contributed to this process, but particularly to the Masters students in Holistic Science who participated in explorations and discussions on the topics covered in the book. I also gratefully acknowledge help from colleagues at the College, especially Stephan Harding, and support from Hazel Goodwin. There are also many others whose insights I have used as inspiration on the journey, too many to name. Thanks to Laura Batson for allowing me to use the curving figures that precede each chapter. Her work can be seen on her web site at curvelanguage.com. Finally, I much appreciate the help I have received with the detailed organization of the book from Christopher Moore, my editor at Floris Books. Thank you all.

Brian Goodwin
25 March 1931 – 15 July 2009

General Introduction

There are moments in the development of cultures when a window suddenly opens on to quite new possibilities that arise unexpectedly from within the culture itself, often in a time of apparent darkness and difficulty. One such moment in Western culture occurred when Europe was emerging from the devastation of the Thirty Years' War (1618–48), which left in its wake uncertainty and confusion about many previous sources of authority and stability, including the Church, the Monarchy, and the classical sources of understanding such as Hippocrates and Aristotle. It was in this period that a window suddenly opened on to a new method of exploring the world. The methodological and intellectual groundwork, laid during the sixteenth century by Galileo and Francis Bacon, then consolidated by Descartes, became the basis of the cultural movement we know as Modernity, giving people a way of exploring the natural world that we call Science. This has been one of the most exciting and powerful adventures in the acquisition of knowledge and power that the world has known. However, we have reached the limitations of this way of knowing and relating to the world and are plunging into another period of darkness and difficulty. This we are experiencing as severe disturbance to the systems on which the quality of our lives depend: the forests, the oceans, the climate system, communities, and cultures, with a devastating extinction rate of species and of human languages, which means also extinction of the associated cultures. This experience is not restricted to Europe, but is now worldwide.

No-one can know what the future has in store for us, just as no-one could have known in 1600 that Modernity was about to flower and take first Europe, and then the world, in a new direction full of possibilities and difficulties. The most we can do is try to read the present signs and to take what seems to be the appropriate action. In this book I read the signs as beckoning in a new direction that has resonances with what came before scientific modernism (that is, the Renaissance, 1450–1600), just as science had resonances with scientific developments in the feudal age (1100–1350) of Roger Bacon, Peter Abelard and William of Occam. The shift of perspective that science could now take arises directly from within science itself so that the transition has recognizable elements of continuity, building on where we have been and what we know. The new direction has the potential to heal many of the separations and divisions that have arisen from ways of understanding nature that were once sensible and useful but have now become too limited to deal effectively with the problems we face. The opportunity presents itself of breaking out of these limitations and expanding our ways of knowing in a manner that encompasses broader perspectives, including the ways of knowing of other cultures.

Nature and culture

The theme of this book is rethinking and extending scientific insight in such a way that nature and culture are understood to be one continuous and unified creative process, not two domains that are distinguished by unique human attributes. Distinctions there certainly are, between non-living and living nature, and between each of the many species of organism that have come into existence on earth. However, I wish to show that the whole sweep of cosmic evolution is characterized by properties similar to those that we jealously regard as our unique attributes, which we use to define the distinction between nature and culture. It has turned out we do not have the upper hand regarding intelligent behaviour; rather, that position is held by nature, which has turned out to be not 'simply mechanical' as our current scientific culture encourages us to believe.

The links between nature and culture that I shall explore throughout the book, in different chapters in different ways, revolve around coherence, wholeness, and meaning. Coherence and wholeness we use to describe natural phenomena, such as lasers and living organisms, but meaning tends to be reserved for humans. I shall develop the position that these three concepts belong together and describe similar processes of creativity in different realms. Our unique attributes of language and consciousness are regarded as the portals that give us access to understanding the meaning of life and of our universe. I shall suggest that looking for the meaning of life is a distracting chimaera, while what we are actually looking for is lives of meaning through relationship. I shall indicate how nature is engaged in a similar process of finding meaning, coherence and wholeness in relationship, and that this is the basis of its intelligibility to us. Meaning, in fact, permeates the creative cosmos that we know.

I shall use a story that comes from another culture to begin this exploration of a possible transformation of understanding that could address many of the difficulties and uncertainties that oppress us at the moment. This is a Zulu story about the African/Asian species of bird known in Zulu as Ngede, the Honeyguide, that leads humans and Honeybadgers to bee hives full of honey. The implications of this story for proper human relationship with nature define the theme of the book.

The Honeyguide's revenge

The children sat before the fire, slowly licking their fingers for the last of the sticky sweetness.

'Ah, Sibonelo!' Gogo smiled. 'You are a good one for finding a ripe hive! We shall have honey at least until the new moon!'

Sibonelo grinned back at his granny. 'It was easy, Gogo! I just followed the Honeyguide.'

Gogo looked at him thoughtfully. 'I hope you remembered to leave the little bird her portion!'

'Oh, yes, Gogo! I would never think of cheating Ngede out of her share!' Sibonelo knew that the Honeyguide would search for a human helper whenever she found a hive that was ready for harvest. While Honeyguide did not care for the honey, she loved to eat the bee grubs and wax from the nest. 'I remember what happened to Gingile, the greedy one, when he took all the honey for himself!'

'What happened to Gingile, Gogo?' asked some of the younger children who had not heard or had forgotten the story. Now that their tummies were full, it was time to satisfy the soul.

'Alright, my children,' laughed Gogo. 'I think a story about little Ngede is appropriate after feasting upon honey she helped bring to our table!' She took a deep breath and began: *'Kwasuka sukela ...'*

There once was a greedy young man named Gingile. He rarely shared with anyone, preferring to keep the meat from any of his kills to himself, hoarding every mealie pip that grew in his small garden.

One day while Gingile was out hunting he heard the honey call of Ngede. Gingile's mouth began to water at the thought of the sweet treat. He stopped and listened carefully, searching until he found the little bird among the branches above his head. 'Chitik-chitik-chitik,' the bird rattled, like the sound of a matchbox shaken lengthwise. When Ngede saw that she had an interested partner, she quickly began moving through the branches toward the nest. 'Chitik-chitik-chitik,' she continued, stopping several times to be sure that Gingile followed.

After thirty minutes or so they reached a huge wild fig tree. Ngede hopped about madly among the branches. She then settled on one branch and cocked her head, looking at Gingile as if to say, 'Here it is! Come now! What is taking you so long?' Gingile couldn't see anything from his place on the forest floor, but he knew Honeyguide's reputation for finding big, ripe nests flowing with sweet honey. He then gathered some dry twigs and made a small fire. As soon as the flames were well established, Gingile put a long dry stick into the heart of the fire. This wood was especially known to make lots of smoke while it burned. As soon as he was sure it was properly burning, he began climbing, the cool end of the branch clamped in his mouth.

Soon he could hear the loud buzzing of the busy bees. 'Ah,' he thought to himself, ' I can almost smell the sweetness in the air. How I love the taste of honey!' When he reached the place of the hive he quickly thrust the burning, smoking end of the branch into the hollow. The bees came rushing out, angry and mean. When most of them were out, Gingile pushed his hands into the nest. He took out handfuls of the heavy comb, dripping with rich honey and full of fat, white grubs. He ignored the few stings he received, placing the comb carefully in the pouch he wore around his chest. When the nest was empty, Gingile slowly made his way back down the tree.

Ngede watched all of this activity with a great deal of antici-pation. She fidgeted nervously, waiting for the moment when Gingile would walk once again on the forest floor and leave, as was the custom, a fat piece of honeycomb as a thank-offering to the Honeyguide. Ngede loved the juicy larval bees and the waxy comb. She flittered from branch to branch, closer and closer to the ground. Finally Gingile reached the forest floor. Ngede flew to a rocky perch near the man and patiently waited for her share. But Gingile put out the fire, picked up his tools and started walking home, obviously ignoring the little bird. Ngede chirped indignantly. She flew before Gingile and landed on a rock in front of the hunter. There she faced the man and crossly called in a high-pitched voice, 'VIC-torr! VIC-torr!' Gingile stopped, stared at the little bird and laughed aloud. 'You want some of the spoils? Why should I share any of this lovely honey with you, you little nothing? Be off and find yourself another sup-per!' And with a wave of his arm in dismissal, Gingile set off for his homestead.

Ngede was furious! How dare this man break the long-time custom and refuse to show his gratitude! But little Ngede was not powerless. She would get her satisfaction! Ngede waited and watched the man for several moons before she sought her revenge.

One day several weeks later Gingile heard the honey call of Ngede. Remembering how sweet and wonderful the last harvest had been, Gingile eagerly followed the little bird once again. After making his way around the edge of the forest, Ngede suddenly stopped her characteristic 'Chitik-chitik-chi-tik,' and came to rest in a great umbrella thorn. 'Ahh,' thought Gingile. 'The hive must be in this tree.' He quickly made his small fire and began his ascent, the smouldering branch in his teeth. Ngede sat and watched.

Gingile climbed, wondering why he didn't hear the usual buzzing. 'Perhaps the nest is deep in the tree,' he thought to himself. He was concentrating so much on his climbing, and was daydreaming about the sweet taste of honey, when he found himself face-to-face with a leopard. Poor leopard was taking her usual midday nap in her favorite tree, exhausted after a long night of hunting, when she was suddenly awakened by a scream. Leopard was first startled and then angry at having her sleep so rudely interrupted. She narrowed her eyes, opened her mouth to reveal her very large and very sharp teeth and took a quick swipe at the man, raking her claws across his forehead. Gingile rushed down the tree, half-falling. He landed with a heavy thud on the ground, breaking several of his bones. Lucky for him that Leopard was still so tired, or she might have decided to pursue the man. Nevertheless Gingile departed as fast as his broken bones would allow him. And he wore the scars of Leopard on his forehead the rest of his life.

Ngede had his revenge, and Gingile never followed a Honeyguide again. But the children of Gingile, and the children of the children of Gingile, heard the story of Ngede and had respect for the little bird. Whenever they harvest honey, they are sure to leave the biggest part of the comb with the juiciest grubs for Ngede!

This wonderful little story is rich in understanding and insight into the meaning of what we learn from nature. In traditional stories about the way the Honeyguide gets what it wants from the bees, there are other players as well. When the Honeyguide discovers a ripe hive, it goes in search of a Honeybadger or ratel that loves honey as much as we do, attracting its attention just as it does with humans, and leading it to the hive. These animals are strikingly coloured, jet black except for a grey mantle that runs down its back from the top of the head to the tip of its tail. With their strong claws and muscular limbs, used for digging and burrowing as do European badgers, the ratel is well equipped to dig into the bee hive, while its fur protects it from the fury of the bees. In the process of getting into the hive and devouring the honey, the ratel scatters lots of honeycomb and grubs for the Honeyguide to eat, so Ngede always gets her share in the process.

I received more information about Honeyguides and Honeybadgers from George Tamanikwa, a native of Zimbabwe who has often seen them in the forests of his native land. He told me that it is important to tell Ngede that you are following her by whistling periodically. If she gets too far away and doesn't hear your whistle, then she comes back to guide you to the hive. The Honeybadger does the same, but instead of whistling, he makes a sort of hooting sound so that Ngede knows he's following. George also told me that you have to be alert when following Ngede because she hates snakes, which steal her eggs and eat her young. So if there is a snake around she is quite likely to lead you to it so that you can scare it away or kill it and protect Ngede's children. Clever Honeyguide! And George gave me one more insight into the intelligence and cunning of the Honeybadger. As described in the story, humans use a smoking stick to make the bees drowsy. When George as a child asked the elders of his community what the ratel uses, they smiled and said that, from what they had heard, he goes up to the hive and farts into it, so that the bees nearly pass out. Clever Honeybadger!

The Honeyguide has another trick up its sleeve. It doesn't build a nest of its own but lays its eggs in the nests of hole-nesting birds, usually barbets. When the Honeyguide chick hatches it usually kills the host's offspring and the host raises the Honeyguide, unaware that it is not one of its own brood. The life of the Honeyguide clearly involves a highly complex web of relationships between different species involving a seemingly endless network of ramifications. We can see that there appear to be winners and losers in these interactions. The bees seem to come out on the short end of the stick after their hive has been raided by humans or ratels. However, their potential for forming new colonies is such that they simply get on with it and soon have a busy buzzing cooperative of their own, with equally complex relationships between its members and the surrounding countryside. The bees forage for nectar and pollen from the flowering plants, returning to the hive and telling their fellows in the colony in what direction to travel to find the food, how far away it is, and its quality. This is all conveyed by the famous waggle dance of bees returning to the hive, and by the nature of the nectar or pollen they bring back with them.

The beehive has long been a symbol of cooperative behaviour in nature that is used to illustrate the emergence of collective order that balances the needs and activities of individuals with the coherence of the colony as a self-organized whole. It is recognized that the waggle dance conveys meaning about foraging sites in symbolic form from individuals who have discovered a new pollen and nectar source to others in the colony, just as we do in sharing useful knowledge with our colleagues (Seeley, 1995). However, we make a sharp distinction between non-human nature and human culture by distinguishing ourselves as the only animals with highly-evolved spoken languages that are fully symbolic and have written forms. We identify this closely with our consciousness and our intelligence, however we seek to describe it. It is this sharp distinction that I shall explore in this book as a basic misconception that underlies much of our alienation from the natural world and the loss of meaning from which we suffer. Separating ourselves from what we call nature by describing it as an objective, mechanical world governed by laws that we discover has allowed us to develop ingenious innovations and inventions in science and technology. However, this form of creativity has confronted us with our awesome power to disturb natural cycles on the planet in ways that now threaten our cultural way of life, due to rapidly disappearing energy, water, and food sources on which we are dependent. This is not very intelligent behaviour. However, it is perfectly natural and can be understood as an aspect of the learning process that we need to go through. Probably all new species had to undergo such a process of discovering whether and how it can fit into the complex network of interacting agents that constitute life on this planet. As a species we are young, and also rather foolish in our current behaviour, which is an aspect of our creativity.

A learning journey

Recently I had an unexpected opportunity to experience directly the significance of relationship as a fundamental aspect of finding a life of meaning. It was the last day of lectures for the autumn term in 2001 and I was lecturing to a group of students at Schumacher College taking the MSc in Holistic Science. Our topic was the dynamics of the heart and the paradox that too much order can be a sign of danger. In the context of our daily lives, or in our experience of working in an organization, the realization that too much order is a sign of danger is commonplace, common sense, so it should not be surprising to find that this holds for the body as well. In Holistic Science we recognize that personal experience comes first and consensus about reliable ways of knowing and its results arise from this through systematic methodologies. So personal experiences and the stories that come from them are seen as the foundations of our connections with the natural world and with each other, giving us important insights into appropriate patterns of behaviour.

In my lecture I was describing the process of ventricular fibrillation and sudden cardiac arrest, illustrating this by means of an electrocardiogram of a subject who was wearing a perambulatory ECG recorder that showed the transition from a relatively normal heartbeat to the final quiverings of the heart as the unfortunate person died . The major part of the recording reveals a very regular periodicity of action potentials in the heart that are running about five times the normal rate (5 per second instead of 1 per second), which is characteristic of a fibrillating heart and is more orderly than the normal heartbeat. This is then followed by a series of smaller amplitude potential changes with much more irregularity, which finally die out as the heart stops functioning due to lack of blood supply and oxygen.

The regularly periodic, rapid action potentials are now understood to be due to a site of repeated depolarizations (electrical action potentials) that can arise anywhere in the ventricle simply from the fact that the muscular tissue of the heart is excitable; that is, it responds to an electrical signal of sufficient strength by depolarizing and then repolarizing, just like a nerve. Therefore any part of the heart is capable of producing a series of action potentials that behave like a pacemaker, initiating rapid heart contractions. This normally doesn't happen because of many specialized aspects of the way the heart generates and propagates the normal action potentials from the heart pacemaker (the sino-atrial node). However, there are conditions under which ventricular fibrillation is facilitated, and this is always very dangerous.

As I was explaining these aspects of cardiac dynamics to the class, I began to feel a bit odd myself, experiencing some irregularities of heartbeat and weakness. I said I was feeling strange, sat down and put my head between my knees. Soon I felt better and looked at the class, who were eyeing me with some puzzlement. I could see that they thought I might be creating a bit of drama to emphasize the points about heart dynamics and the fact that these things can happen to anyone, anytime, though they are normally very improbable. I said that there was only one other topic I had intended to talk about that day: the evidence for chaos in the dynamics of the brain. However, in view of what was happening to me as a result of talking about ventricular fibrillation, I decided not to invite the risk of a chaotic mental state. We ended the seminar and I went and lay down.

The symptoms didn't go away, so we decided that I'd better get checked at the local hospital in Totnes, Devon. The examining doctor realized that something was wrong, but it wasn't clear what the trouble was. So I went by ambulance to another nearby hospital (Torquay) with better facilities, and there they discovered that I was undergoing what in the trade is called a dissecting aneurism of the aorta. That is, the major artery from my heart was peeling apart, resulting in irregular blood flow. I wasn't fibrillating, but the condition requires emergency treatment. So then it was 'Blue Alert' in an ambulance to the hospital in Plymouth where an expert cardiac surgical team prepared to operate. The chances of surviving an aortic aneurism are only 5%, because the condition doesn't give clear symptoms and it is usually too late by the time a diagnosis is made. I was one of the lucky ones. I remember very little of what happened after leaving the hospital in Totnes because I was on morphine for the journey to Plymouth, then out on the operating table a few times, followed by several days under sedation in intensive care.

The surgical team replaced my aorta with one from a cow, and my aortic valve with one from a pig, to whom I am deeply grateful for my multi-species nature. These tissues have no living cells in them, so my body accepted them without any immunological rejection. The operation was long but successful and I returned to intensive care, where my breathing and heartbeat were both driven by external control, overriding my own pacemakers. There were complications and I went back into surgery twice, each time returning to intensive care where I wobbled between this world and the next. It was during this time, when my body was making up its mind which way to go, that I went on some remarkable trips, some of which I have come to regard as learning journeys while others were funny, bizarre, or fascinating, but all were intensely vivid. When I began to come round again, I was convinced that they had really happened. Here is one of the learning journeys that was about opening the heart and love.

While sedated in intensive care, so that according to normal criteria I had no signs of conscious awareness, my psyche was extremely active and I was experiencing intense visions and emotions. I found myself in a landscape like the foothills of the Rockies, familiar to me from my many visits to Santa Fe over the years. The rising mountains were covered with aspens whose leaves are capable of the most beautiful quivering motion. I raised my right hand and rotated it slightly. In response, the aspen trees started to sway back and forth and the leaves quivered with feeling. I then noticed that there were a lot of animals among the trees. There were not only deer and antelope, but also African savannah animals like Thompson gazelles and springbok. I then rotated my forearm around, and in response the animals began to leap about and turn somersaults. I was totally enchanted by this spectacle, filled with an extraordinary sense of peace and connection to the plants and animals.

Then I heard a voice which said: 'It's all very well being connected like this to plants and animals. However, can you raise yourself up to the top of that mountain on the power of love alone?' I had no idea how I might achieve that, but I knew it was a test of my worthiness to live, or to die. I tried to raise myself up, but got nowhere. Then I began to call on students, friends and colleagues from the College, to help me. They came from everywhere: from Alaska, from Montana, from Canada, from New Zealand, from Germany, from the UK and many other countries, some on skis and some on foot. We had no idea what to do, but suddenly we had the inspiration to move in a circle together and create a vortex. This vortex became a vivid pink-magenta colour and began to rise up, carrying us effortlessly to the top of the mountain. But whenever I found myself separated from them, I felt myself slipping down again. So it's easy to get to the top of the mountain, but you can't do it on your own. You need to be carried up by the collective power of love from the community to which you belong. I realized later that my heart had been opened literally and I had survived through surgery. But the more important message was to live with an open heart. This is what Holistic Living is about.

Science with meaning

The science of the twentieth century has ushered in new ways of understanding and knowing the world that belong within science but transform the boundaries usually given to scientific inquiry. I explore these in successive chapters of this book. Chapter 1 describes first the historical

context from which modern science and the mechanical worldview arose;
then how the life sciences have transformed under the impact of chaos and
complexity theories so that emergent properties are recognized as the stuff
of evolution, both physical and biological. Furthermore these properties
are intrinsically unpredictable, though they are intelligible in retrospect
so they can be 'explained' after the event. We are thus confronted with an
uncontrollable and creative but self-consistent and intelligible reality.

The implications of this new perspective in biology, particularly
regarding the health of individuals, communities and ecosystems, is
explored in Chapter 2. Here the new view of health as a robust dynamic
process with chaos as a creative component is described for basic physiological processes and in connection with cancer. What emerges is then
the importance of meaning in relation to healing, the recovery of wholeness and coherence. This carries over into ecosystems and their use in
food production, giving us an insight into the dangers of genetic manipulation and its effects on the health of farming communities.

Chapter 3 examines the shift from a science of quantities to a science
of qualities under the impact of the new thinking about how we come
to know and to understand the emergent properties of complex systems.
This major transformation of science is examined for its implications
regarding the status of qualities and values in nature.

The consequences of these shifts in science regarding our understanding of biological evolution are described in Chapter 4. Here meaning
enters the dynamics of living process in describing how organisms make
sense of the historical information bequeathed to them in their genes by
creating themselves as forms that express embodied meaning. Language
is involved in this process in terms of the self-referential networks within
cells that create forms with meaning from genetic texts.

In Chapter 5 the ideas of self-similarity and fractal structure, introduced in connection with chaos and complexity in biology, are described
in physical processes as similar expressions of creative dynamics in
natural processes. These reveal generic features of living and non-living
activities when forms are coming into being. It emerges that the natural
world shares with us the same type of creative process as that which we
experience in culture, so that nature and culture become one.

An early expression in European culture of this same theme is
described in Chapter 6 through the work of Johann Wolfgang von
Goethe. His writings contributed significantly to the movement that in
literature became known as nineteenth century Romanticism, while in

the twentieth century Goethe's scientific method has been seen by philosophers as a forerunner of phenomenology. This allows an extended science to embrace both quantities and qualities as necessary descriptors of the creation of natural forms, uniting science with the arts and humanities for effective education and praxis in our present era.

Chapter 7 examines the ways in which we can engage with the Great Work of cultural transformation that arises both from within, as a natural historical process, and from the external necessity to adapt to the new situation we find ourselves in as our familiar ecosystems collapse and climate changes. The new science described in previous chapters then becomes the conceptual and methodological foundation for the practical work of re-educating and re-equipping ourselves for the transition to a new lifestyle of cooperation with other cultures and with nature, all understood in the same fundamental terms as creative enterprises.

Science evolves in interesting and unexpected ways, like everything else in nature. In the twentieth century the natural dialectic of scientific discovery and understanding underwent a series of dramatic transformations that have now resulted in a window opening on to a richer and more meaningful way of practising science and participating with the world than we have been engaged in. Questions about the relationships between facts and values and proper conduct in relation to the nature that we have come to understand are now reappearing on the scientific agenda in challenging and urgent ways, after being outlawed from the scientific agenda for four centuries. The sharp distinction that we have made between nature and culture is beginning to dissolve, leaving us with a more unified but also a more complex world in which to pursue our lives. Unless we want to end up with broken bones and the mark of the angry leopard on our foreheads, like Gingile, it is time to expand our understanding, mend our manners and learn to share the feast with the Honeyguide and all the other participants in our planetary community. This is the challenge to which the union of science with the arts through natural creativity, explored in this book, is a response.

1. From Love to Gravity

At a conference in the stately and beautiful Villa Serbelloni on the shores of Lake Como in Italy, I vividly remember a paper presented by an American historian and philosopher of science, Steven Goldman, in the early 1980s. I had been to the Villa several times before for a series of conferences organized by the British biologist C.H. Waddington, my PhD supervisor, on the topic 'Towards a Theoretical Biology.' To the last of these four conferences I contributed a paper entitled 'Biology and Meaning,' which was the final paper of volume 4 (Waddington 1972), the last of the publications from the series. In this article I wrote:

> Thus science as we know it has largely opted to pursue the course of manipulation and power, drawing us inevitably into a Faustian crisis which arises from the irreconcilability of manipulation and wisdom. To manipulate wisely one must be wiser than Nature, wiser than Man, for both must be manipulated; hence one must be God. Faust found that only the Devil would play this game with him, tempt this hubris. The corruption arises with the decision to manipulate rather then to engage in a dialogue, the decision to be Master rather than partner.
>
> What, then, is the alternative to the dominant contemporary attitude in science? How can we redefine the scientific enterprise so that it is coupled with the search for meaning and wisdom, not just the acquisition of knowledge and

power? In what way is it possible for us to participate in
this enterprise of learning about the world?

Steven Goldman's paper, entitled 'From Love to Gravity,' presented ten
years later in the same room where I had asked these questions, helped me
to understand how and why science had arrived at this point. He discussed
the emergence of modern science from Renaissance nature philosophy that
prevailed in Europe during the period from about 1450 to 1600. The view
of the cosmos at that time was vividly presented in the writings of the magi,
the wise men, such as Marsilio Ficino, Pico della Mirandola and Francesco
Giorgi in Italy, Cornelius Agrippa and Johannes Reuchlin in Germany,
and John Dee in England. They described reality as a single coordinated
domain, every region of which is intrinsically related to every other region,
so that to know the region called nature entailed knowing also the nature
of the being that knows. The resulting union of the knower and the known
meant that a change in one resulted in a co-ordinated change in the other,
mind and nature undergoing cooperative transformation during the process of real learning in a world that is alive with agency, with creativity
and intrinsic value. The cosmos of the Renaissance magi was composed
of intrinsically correlated elements held together by a force called love.
Therefore, to achieve knowledge of nature required perfection of one's own
physical, moral and spiritual, no less than intellectual, characteristics.

This is the world that comes vividly alive in the works of Shakespeare,
who describes both the tragedies and the glories of the human drama of
participation in a world of magic, music, love and harmony. In *The
Merchant of Venice,* the Christian Lorenzo, captivated by the beauty of
the night, tells his love, the Jewess Jessica, about universal harmony:

> How sweet the moonlight sleeps upon this bank!
> Here we will sit and let the sounds of music
> Creep in our ears: soft stillness and the night
> Become the touches of sweet harmony.
> Sit, Jessica. Look how the floor of heaven
> Is thick inlaid with patens of bright gold:
> There's not the smallest orb that thou behold'st
> But in his motion like an angel sings,
> Still quiring to the young-eye'd cherubims:
> Such harmony is in immortal souls,
> But whilst this muddy vesture of decay
> Doth grossly close it in, we cannot hear it.

In striking contrast with the vision of the magi and Shakespeare's world of beings united by love, the personal and especially the moral qualities of the seeker after knowledge in modern science are seen as at best irrelevant and are more generally regarded as an obstacle to achieving knowledge. Facts and values need to be rigorously separated so they don't contaminate each other. This follows from the depersonalization of the pursuit and achievement of knowledge in the Western scientific tradition, which sees reality as composed of atomic constituents that are only extrinsically correlated with one another, obey generic laws of interaction that make no provision for individual characteristics, and are held together by a mysterious force called gravity. So love turned into gravity and the place of creative personal agency and qualities in nature disappeared. For the Renaissance magi, the world was intelligible, creative and meaningful. For the contemporary scientist, it is intelligible only, having no meaning in itself.

Going back to the early Renaissance and trying to extract insights relevant to understanding our present position is a difficult and uncertain process. We simply weren't there to participate in the context we seek to understand and make sense of in relation to where we are now. The point is not to dwell very long on these issues but to take a message from them that may help us to follow the dialectical unfolding that is now happening before our very eyes, as the simplistic but powerful and undoubtedly effective worldview of 'science' impales itself on its own contradictions. This is what living, dynamic traditions do to themselves: they transform in unexpected and remarkable ways. The limitations of the magian, pre-modern worldview are very clear to us and there is no question of trying to go back there to recover a lost world. Knowledge was transmitted largely through authorities, whether institutional (like the Church) or individual (like the magi themselves). The Church tended to keep within its own authority the key to individual transformation, the possibility of removing Shakespeare's 'muddy vesture of decay,' by adherence to Church Doctrine which was the only route to personal salvation. The magi disputed this, claiming that a person can undergo transformation through redemption by gaining knowledge of reality, which must involve personal purification. This was revolutionary stuff, elevating the person above the church, and the magi paid a price for such claims: Giordano Bruno was burned at the stake in 1600 by Church authorities for his heretical views on the potential of the individual to achieve enlightenment and power reflecting that of God through knowledge and transformation.

The modern scientific movement, subsequent to the pioneering work of Galileo, showed how any individual can indeed gain reliable knowledge of nature by following the scientific method of experimentation, measurement, and discovering the mathematical relationships that capture the patterns of dependence between measurable quantities, described as the laws of nature. The basic convictions that underlie the possibility of approaching the world in this manner were established by medieval scientists such as Peter Abelard, Robert Grosseteste, Roger Bacon and Adelard of Bath, in the twelfth and thirteenth centuries. They argued that knowledge of nature does not require reference to extra-natural powers and beings. It depends on the application of human reason in the form of mathematics applied to natural phenomena and the cultivation of an impersonal experimental method of observation. These principles re-emerged as the foundation of the scientific method developed by Galileo and elaborated on by Descartes and Francis Bacon in the sixteenth and seventeenth centuries. The truly liberating aspect of this method of knowing was, I believe, not the experimental study of quantities and the mathematization of nature, powerful as this is. Rather, it is the procedure whereby individuals reach consensus about their knowledge by engaging in a community of practice. In a scientific community there are no authorities. Each individual is free to pursue his own agenda of work, and to disagree with the majority if the work points in this direction. However, the person must then convince other scientists that his interpretation of the observed phenomena is more consistent with the evidence than the prevailing view. The process of reaching consensus is subtle and complex, as it does not depend on majority voting or on authority. It depends entirely on consistency of interpretation and the unity of scientific understanding achieved.

Let's now take a look at the warning bells that have gone off as science has pursued its goal of reliable knowledge according to the depersonalized approach to an independent reality composed of independent entities. What we need to remember is that the knowledge that has come from this approach is powerfully useful and effective, and its relevance for the creative transformation of aspects of our world remains undiminished. The way of understanding the world that came from the insights of the founders of modern science such as Galileo, Descartes, Francis Bacon and Isaac Newton is a glorious achievement that must not be lost or discarded in any new paradigm shift that might occur. However, its limitations are now beginning to overwhelm its effectiveness. We need to understand why, and how these limitations may be overcome.

Before I go into the changes that are challenging Western science from within and without, let me illustrate what is happening with a modern fairy tale that tells the whole story in a nutshell. The tale comes from John Fowles' novel *The Magus,* published in 1966, a time when the whole of Western culture underwent a sudden spasm of expansion and opening. The hero of the novel is a young man who undergoes a learning process, organized by a modern magus who reflected the values of earlier magi, focussing on the importance of inner awakening to the complexities of the human psyche involved in living fully the qualities of experience.

The prince and the magician

Once upon a time there was a young prince, who believed in all things but three. He did not believe in princesses, he did not believe in islands, and he did not believe in God. His father, the king, told him that such things did not exist. As there were no princesses or islands in his father's domains, and no sign of God, the young man believed his father.

But then, one day, the prince ran away from his palace. He came to the next land. There, to his astonishment, from every coast he saw islands, and on these islands, strange and troubling creatures whom he dared not name. As he was searching for a boat, a man in full evening dress approached him along the shore.

'Are those real islands?' asked the young prince.

'Of course they are real islands,' said the man in evening dress.

'And those strange and troubling creatures?'

'They are all genuine and authentic princesses.'

'Then God also must exist!' cried the prince.

'I am God,' replied the man in full evening dress, with a bow. The young prince returned home as quickly as he could.

'So you are back,' said his father, the king,

'I have seen islands, I have seen princesses, I have seen God,' said the prince reproachfully.

The king was unmoved. 'Neither real islands, nor real princesses, nor a real God, exist.'

'I saw them!'

'Tell me how God was dressed.'

'God was in full evening dress.'

'Were the sleeves of his coat rolled back?' The prince remembered that they had been. The king smiled.

'That is the uniform of a magician. You have been deceived.'

At this, the prince returned to the next land, and went to the same shore, where once again he came upon the man in full evening dress.

'My father the king has told me who you are,' said the young prince indignantly. 'You deceived me last time, but not again. Now I know that those are not real islands and real princesses, because you are a magician.'

The man on the shore smiled. 'It is you who are deceived, my boy. In your father's kingdom there are many islands and many princesses. But you are under your father's spell, so you cannot see them.'

The prince returned pensively home. When he saw his father, he looked him in the eyes. 'Father, is it true that you are not a real king, but only a magician?'

The king smiled, and rolled back his sleeves. 'Yes, my son, I am only a magician.'

'Then the man on the shore was God.'

'The man on the shore was another magician.'

'I must know the real truth, the truth beyond magic.'

'There is no truth beyond magic,' said the king.

The prince was full of sadness. He said, 'I will kill myself.'

The king by magic caused death to appear. Death stood in the door and beckoned to the prince. The prince shuddered. He remembered the beautiful but unreal islands and the unreal but beautiful princesses. 'Very well,' he said. 'I can bear it.'

'You see, my son,' said the king, ' you too now begin to be a magician.'

Western science, like the young prince, arrived during the twentieth century at a land where things were not as expected from previous scientific knowledge about the world. For three centuries, from the seventeenth to the twentieth century, scientists dedicated their lives to discovering the real truth about nature, interpreted as law-governed processes that can be understood in terms of mechanical causality, resulting in prediction and control of many of these processes. This was a stunningly effective way of getting reliable knowledge about many aspects of reality that can be used for a variety of technological processes. These include methods of food production, ways of combatting disease by controlling infections, electromagnetic devices such as radio, TV and the internet for communication, ways of amplifying power through a variety of machines that allow us to extract resources from the earth and to travel around it, to name but a few. However, it has now become evident that this form of knowledge and power applies to only limited aspects of a reality that turns out to be, in general, holistic, unpredictable and creative, for reasons that have become clear through developments in science itself .

A major result of this dialectical change in our culture is that we have to learn how to live in this new world in a different way, cultivating new skills and enlarging our scientific horizons so that there is room for the creativity that is a part of our everyday experience of living.

When I started my scientific career in the 1960s, there were three major taboos that defined the boundary between what was scientific and therefore defined reality, and unscientific territory, the unreal. These were consciousness, qualities, and animism. Consciousness was something you couldn't talk about seriously as a scientist because it was associated with subjective experience, and so was not accessible to scientific study. Qualities such as the experience of happiness, grief, trust or love were also identified with subjectivity, the cloudy and unreliable world of personal experience where scientific investigation could not gain a foothold. While these were acknowledged to have some kind of status in experience, animism, the belief that rocks or trees or crystals themselves have feelings, was regarded as evidence of a serious deficiency in the individual's ability to reason about the world. Like the prince in the fairy tale, instructed by his father, I accepted these taboos unthinkingly as the wisdom of my scientific superiors. However, now consciousness is definitely on the scientific agenda, qualities are emerging in various areas of scientific study, and animism is on the horizon, though still out of bounds for most scientists. Western science has now developed to the point where, like the young prince, scientists are finding it necessary to get beyond the desire for certainty and live with their new understanding of reality by becoming participants in the unexpected qualities of natural creativity. They are becoming part of the unpredictable magic of natural process, but without losing the essence of the scientific way of gaining reliable knowledge through consensus in a community of practice.

The transition from mechanistic to holistic science

The first awakening of western science to the unexpected subtlety of the world came in the early years of the twentieth century with Einstein's discovery of relativity as the way to describe the relationships between different observers in a world where communication is not instantaneous but is limited by the velocity of light. There is no absolute frame of reference, no preferred perspective that gives one observer authority over another in observing the unfolding of natural processes according to intrinsic laws of motion of gravitating bodies, the stately march of the planets or other aggregations of matter of which the cosmos is composed. Each observer is free to choose whatever frame of reference is most convenient and elegant for describing the processes being observed, and consistency with other observers' chosen reference frames depends upon mathematical relations that are well-defined.

The second wake-up call to occur in science was the development of quantum mechanics. The essential point for my story is that quantum mechanics reveals a physical reality that is holistic. The world does not consist of independent particles whose characteristics of position, momentum, electrical charge, spin, and so on, can be varied independently of each other. The quantum realm is governed by principles of intimate entanglement and coordination between its components, a non-local connectedness resulting in holistic, correlated order that extends over any distance. This is a far cry from the mechanical universe of the nineteenth century in which whatever bit of the world you may be studying is just the sum of many independent parts in interaction, like the interacting components of a machine.

However, it was generally assumed by scientists that such strange holistic behaviour was confined to the world of very small elements such as electrons, protons, neutrons, photons and the like. The world with which we are familiar, such as rocks and water, plants and animals, cars and refrigerators and bicycles, was still considered to behave in much more common-sensible ways. They work independently of each other so that we can predict what will happen to them under various circumstances. We also understand them to be made up of smaller parts so that the whole is the sum of these parts in interaction. Different parts can be added or substituted to modify the properties of the whole, as when we add antifreeze to water to change its freezing point, or replace a diseased kidney with a healthy one in a human to restore the whole to a condition of health. This is a very powerful and effective view of the world, and it has given us the vast array of technologies that make our lives considerably easier. However, this view of 'normal' reality is now also coming to an end because we are waking up to its limitations.

Chaos and unpredictability

The first realization that the world of every day experience is not predictable and controllable came in the 1960s, when chaos was discovered to be an intrinsic aspect of the weather. Of course, in a colloquial sense we knew this all along: weather is unpredictable, chaotic. However, it was assumed that this is because we didn't have enough detail about atmospheric states from observation satellites and weather stations, or powerful enough computers to make accurate predictions. Give us the details and the computing power and we'll make predictions as long-term and accurate as you want, said the meteorologists. However, a mathematical discovery put an end to these claims. The discovery was called deterministic chaos: processes described by mathematical equations can behave in ways that are unpredictable to an observer. The equations used to describe the weather have this property. Therefore, for reasons intrinsic to the dynamics of the weather itself, it is unpredictable in the long-term, no matter how good our observations of current weather conditions and how powerful our computers.

The behaviour of mathematically-defined processes can be described by representations of their activity by what are called attractors, which show their dynamic characteristics. For example, a repetitive process such as the motion of a clock pendulum can be represented by a closed curve that describes the repeating cycle in terms of appropriate variables such as the position and velocity of the pendulum at any moment in time.

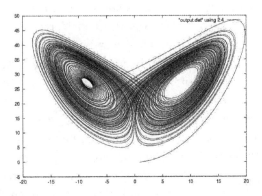

Figure 1. The structure of the Lorenz attractor

A complex process such as the weather revealed a new type of pattern called a strange attractor, in which there is never a repetition of activity although the process stays within certain bounds that contain the attractor. The best-known example of a strange attractor is the Lorenz attractor, shown in Figure 1. This is named after Edward Lorenz, who studied a simplified set of equations describing the behaviour of the weather system in the 1960s at MIT. The pattern of the trajectories that describe the weather describe a subtle kind of order that reflects what is known as a fractal structure to the dynamics: the attractor has a fractional dimension in the space in which it exists, and this property is shared by any part of the attractor so that the part and the whole are self-similar.

Actually, as is so often the case in science, this property of mathematical equations was first discovered many years before the realization about the weather. At the end of the nineteenth century, the French mathematical physicist, Henri Poincaré, showed that the equations describing the motions of the planets could produce deterministic chaos. So planetary motion, long assumed to be the essential manifestation of the orderly dance of the mechanical universe, turned out to have unpredictable components which have recently been confirmed from observations. The motion of Pluto, for example, has chaotic elements, as do the motions of Jupiter and Saturn (Murray and Holman 2001). As a result, there is always the possibility that planetary excursions may exceed a critical value so that a planet gets ejected from the solar system. Planetary motion is not quite so stately as had been thought for millennia, and the stability of our remarkable planetary home is not as certain as we had assumed. In a creative cosmos nothing can be taken for granted, encouraging us to live not in time but in the present moment of creative becoming. The insights that come from the discovery of deterministic chaos, both metaphorical and analytical, are now everywhere in science, including biology. Perhaps its most surprising appearance is in the dynamics of the heart and the window this provides on to the health of the body as a whole, as well as giving us a different way of looking at cancer. These will be major topics of Chapter 2.

As chaos was working its changes on scientific understanding, another development was taking place that stimulated a fourth awakening of science to the unexpected creativity of natural processes. This again started in physics but spread rapidly to biology. Physical systems sometimes behave in quite unexpected and interesting ways, and the task of the physicist is then to try to make sense of these phenomena in terms of simpler properties that are already, to some extent, understood. Physicists studied sudden transitions that materials can undergo, from one type of order to another through disordered states, such as the abrupt freezing of water to form ice. Biologists then recognized that similar types of transition occur in living systems. Individual birds or fish behaving independently of one another can come together and form flocks or schools that move with the most breathtaking coherence, suddenly becoming a unified whole like a single organism. Social insects such as bees, ants and termites do the same. The behaviour of individuals or small groups is often highly disordered, even chaotic in the technical sense of the term, but when many individuals interact they can perform miracles of coordinated activity: termites construct their intricately pillared apartments and passageways, while ants coherently tend the brood and the queen in the nest although there is no one in charge.

Such unexpected, orderly patterns from disorderly elements are known as emergent properties. In general they cannot be predicted from a knowledge of how individuals behave and how they interact. Nevertheless, every time particular conditions of behaviour and interaction occur, the same organized pattern of the whole emerges. It is now recognized that emergent properties are very widespread in nature, particularly in living systems. Many of the most intriguing characteristics of life, such as the way a complex organism emerges from the interaction of many cells during embryonic development, or the pattern of species extinctions during evolution, are unexpected results of particular patterns of interaction between components in complex systems (Solé and Goodwin 2000; Camazine *et al* 2001). The world is now seen to be full of emergent properties, which is the scientist's way of recognizing the creativity of natural processes. These new insights into emergent properties are altering the way in which we attempt to understand and explain natural phenomena, especially in the context of evolution, as I shall describe in Chapter 4.

Emergent phenomena in nature

In the course of exploration and discovery in conventional science, there are occasional moments when it is recognized that nature is behaving in ways that are deeply puzzling to our expectations. An example of this in physics is superconductivity in metals, where the electrical resistance of the metal vanishes below a critical temperature and an electric current will propagate around a loop of the metal indefinitely — a kind of perpetual motion! An example from chemistry is the Beloussov-Zhabotinsky reaction, in which a particular mixture of chemicals spontaneously produces spatial patterns, both concentric circles and spirals, that propagate through the medium. Since these phenomena are well-defined in their behaviour, and repeatable, it is assumed that they can be understood logically. What is often required in these cases is an extension of the principles that operate in more familiar situations. Superconductivity required the introduction of a new type of interaction or coupling between electrons, while the Beloussov-Zhabotinsky phenomenon required the recognition that chemical reactions could be spontaneously periodic, even when the reagents are mixed uniformly in a dish. In all such cases, the phenomenon was not predictable from what was then assumed or known, and assumptions about the underlying constituent processes had to be modified to explain the behaviour.

Biology is full of such unexpected phenomena. We are still asking the question: how did life arise on our planet? We see that living beings modify conditions on earth in dramatic ways, often so as to stabilize conditions such as temperature or the oxygen content of the atmosphere or the concentration of salts in the oceans (Lovelock 1991; Margulis 1999). But the everyday phenomena of life such as the shapes of organisms or the patterns of their behaviour are also properties that emerge from complex interactions of their parts in ways we do not understand. This is where complexity theory is extending the type of insight we have into the dynamics that give rise to novel forms in evolution.

Evolution and emergence

Evolutionary biologists have been struggling with questions about the origins of species diversity and their adaptive properties since it was accepted that species have emerged through natural processes and are not created by a divine, all-powerful being. Darwin assumed that the primary problem in explaining the origin of species is how it is that living organisms are so well adapted to their habitats: worms to the soil, fish to the seas, birds to the air. His answer, as everyone knows, is that there are small, spontaneous, hereditary variations within the members of any species that result in diversity of properties. Those variants that are most suited to the habitat in which the species lives are then selected or stabilized as the fittest adaptations, allowing members with these properties to leave more offspring and so to predominate. Particular species are simply those organisms that are sufficiently well differentiated in their adaptations to be recognized as distinct types, due to the accumulation of a sufficient number of distinguishing adaptations. This is the sense in which species are considered to be created by natural selection: they are a result of an accumulation of adaptations each of which is selected for its fitness, enhancing the capacity of the species to survive in a habitat and to leave offspring. Simple and powerful as it is to account for many adaptive aspects of organisms, this explanatory framework has its limitations in explaining how particular characteristics of species arise in the first place.

It is useful to look at the type of explanation that is accepted for an emergent property in biology in order to clarify these issues about the explanatory power of natural selection in evolution. The example that I shall use for this is the emergence of a coherent rhythm in the brood chamber of certain species of ant from interactions between workers whose individual behaviour is chaotic in the technical sense of the term. The question is: how does a regular periodic rhythm, characteristic of behaviour in certain species of ant and of value in their reproduction, arise from the interactions of chaotic individuals?

It was established from studies by Franks *et al* (1990) that the workers in an ant colony of the genus Leptothorax, tending the queen and the young in the brood chamber, exhibited a well-defined rhythm in their activity-inactivity cycles. Roughly speaking, the workers took a rest for about ten minutes every half hour. The first assumption was that each ant is intrinsically periodic in its activity-inactivity cycle and in the brood chamber these cycles become synchronized through interactions so that the whole exhibits a regular periodic cycle. Such synchronization is well-known in the flashing of fireflies or the coming into oestrus (fertility) of female monkeys with the full moon. However, when Cole (1991) examined the behaviour of individual ants from colonies of a species closely related to that studied by Franks, which also exhibited collective rhythmic behaviour in the brood chamber, he found that the activity-inactivity cycles of individuals were technically chaotic. Single ants, or a few ants at low density, exhibited deterministic chaos without any sign of a regular rhythm. How, then, does a regular periodicity emerge in a colony from the interaction of chaotic individuals?

This problem was studied by Solé *et al* (1993) by constructing a computer model of the interaction of ants in a brood chamber. Each individual model ant was described in its activity-inactivity behaviour as chaotic, in accordance with the experimental evidence, and the model ants interacted with each other by excitation, again in accordance with observation: if an active ant interacts with an active or an inactive ant, both cases result in increased activity. The model was constructed as a kind of mobile neural network, ants playing the role of excitable neurons that can move around on a grid that represents the brood chamber. It was not possible to predict the outcome of the model, since there was no reason to assume that chaotic individuals interacting by excitation will develop a coherent rhythm. However, that is precisely what occurred. Not only did an activity rhythm emerge from the interactions but it conformed to observations that were not built into the model: a regular periodicity emerges in the ant colony only when the density of ants exceeds a critical level (about 20% of maximum density in the colony), as observed experimentally. Furthermore, the regular rhythm emerges quite suddenly above the critical density; in physical terms it has the characteristics of a phase transition, like the sudden condensation of steam to water at a critical temperature and pressure. At the critical density the ants have a distinctive pattern of activity: there are rhythmically organized clusters of ants of all sizes, from two to the full size of the colony, that form transiently and then disperse. These clusters have a characteristic size distribution that follows what is known as a power law. This describes a self-similar or fractal pattern in the dynamics of the colony. At this point it is as if the ants are exploring the range of possible order that can emerge as a coherent pattern, the fractal structure finally condensing into coherent rhythmic order over the whole colony when the critical density is exceeded. We shall find this type of pattern as a signature of the coming-into-being of order in many different systems, both living and non-living. It is now possible to reflect on what kind of explanation of emergent order this model gives us.

Biological form and natural selection

Patterns of behaviour in organisms are instances of biological form, presenting us with dynamic order in time and space. The emergent periodicity in the brood chamber described above cannot be reduced to the behaviour of the individual ants and their interactions in any specific causal sense. That is, knowing about individual ant behaviour and the nature of their interactions is not sufficient to predict the periodic behaviour of the whole colony. Nevertheless, this periodicity is a consistent property of the model, above a critical density of ants. Here is a case of consistency without causal reduction, which is found in many cases of emergence in complex systems. What the model makes clear is the properties required for such an emergent process. The mathematical functions used in the model describe these properties without explaining what their consequences are. This is typical of scientific procedure: models often describe without giving causal explanations of the phenomena.

Models of the type outlined for ant behaviour are very useful in clarifying various aspects of emergent properties in complex evolving systems. In particular, we can ask what role natural selection might play in the emergence of rhythmic activity of workers in the brood chamber. To explore this, we need to ask what function rhythmic activity has for the colony and how it serves its survival. Franks suggested what this might be. He and others have observed that if an active ant encounters an embryo or a larva that is already getting attention from other workers, the active worker will go to another member of the brood. Therefore if workers in the brood chamber are active at the same time, they tend to distribute their care over the brood and the queen so that there is little duplication of attention. If their activity patterns were chaotic, then some embryos or larvae might get more, and others less, attention than others, simply by accident of how active workers are distributed in the brood chamber. Solé has shown that when this is modelled by computer simulation of colony activity, periodic patterns do indeed produce better care than chaotic patterns. So we have a possible explanation of why rhythmic cycles of activity in the brood chamber are beneficial and may be selected: they enhance the survival chances of the young by virtue of good care when they are developing, and so increase the fitness of the colony. This is probably why these activity cycles are commonly observed in such colonies.

However, it is clear that natural selection in no sense explains the *origin* of the rhythmic activity pattern in the brood chamber. The possible function of the rhythm played no role in designing the model, which provides an understanding of how rhythmic activity patterns are possible in an ant colony with the properties observed. Modelling the ant colony as a complex system provided a demonstration of how rhythmic behaviour in the brood chamber could have arisen spontaneously from the interaction of ants. It is an example of self-organization in a complex system as the origin of a biological form. It is clear that any biological form must arise spontaneously before it can be selected, and one of the jobs of science is to provide plausible explanations of how this might occur. Natural selection can explain the differential abundance of different morphologies and behaviours in different species, relating to their utility, but it can never explain the spontaneous origin of biological forms. Darwinism and Neo-Darwinism propose that new forms arise as a result of random change in genes. This may well be the case, but we are then left asking how the observed patterns and forms of organisms are generated from known properties. What makes them possible? Complexity theory addresses this question of origins, providing an explanation by describing a pattern of interactions in a complex system from which the form can arise.

Natural design

Among the forms produced by organisms are their constructions, such as the nests of social insects and beaver dams. These are examples of natural design in operation. A highly instructive example for issues relating to good ecological design comes from the study of termite colony nests, which can reach a height of a few metres in the African savannah. How do they construct such intricate dwellings, with pillared halls, chambers and apartments, complete with air conditioning and temperature regulation? Detailed studies of termite behaviour while they are building these structures reveals that the whole process is a perfect example of self-organization, requiring no architect or boss to design and direct the process. A worker starts building by placing a fecal pellet, a mixture of earth and excreta, at a site. The pellet is impregnated with termite pheromones which attract other workers to the site, to which they add their pellets, working them into the construction. This process is similar to cob construction of houses by humans using a mixture of earth, water and straw as the building material, which dries into a solid structure as does the termite dwelling. As termites add their pellets, more termites are attracted cooperatively and the pillar or roof or other construction grows. Then suddenly the group gathered at a site freezes, work stops, and they all scatter. This happens for groups of any size so it is density of workers, not number, that counts. Individuals then prepare new pellets and disperse to other sites or start new ones and the process of attracting other termites to the construction process happens again. This seemingly disorganized activity has the unexpected consequence that the pillars, walls and roof of the extension to the colony grow coherently, as shown by Myerscough and colleagues (2000) in a model of this process. The combination of attraction of workers to a site by pheromone-loaded pellets, and inhibition of activity at a certain density, results in an even distribution of workers over the construction area, achieved through a fractal pattern of worker numbers at sites and then their random scattering to new sites. As mentioned in connection with the emergence of rhythmic order in the brood chamber of ant colonies, such fractal patterns are frequently found when a form with long-range order or coherence is coming into being, whether in the living or the non-living world. This is a robust and reliable way to generate order on different scales. Termites achieve coherent constructions through use of natural materials and a self-organizing dynamic, a model of sustainable and robust architecture from which we have much to learn.

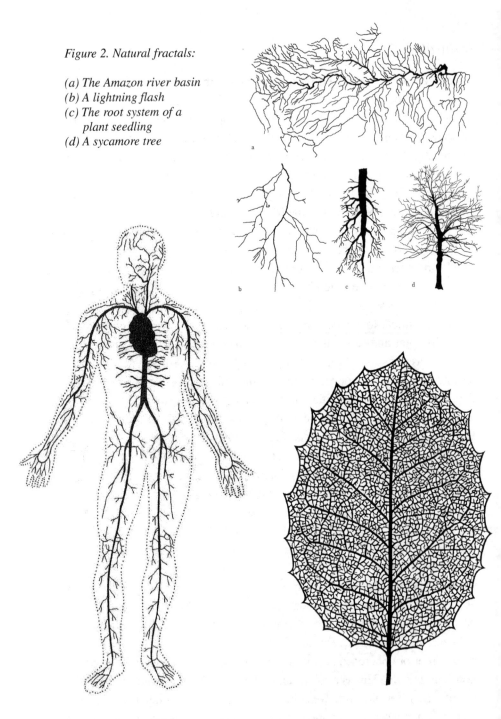

Figure 2. Natural fractals:

(a) The Amazon river basin
(b) A lightning flash
(c) The root system of a
    plant seedling
(d) A sycamore tree

Figure 3. The human circulation

Figure 4. The venation pattern
in a holly leaf

Fractal patterns in space

The most familiar fractal patterns in nature are those extended in space, such as the ones shown in Figure 2: (a) the Amazon river basin, (b) a lightning bolt, (c) the root system of a plant, and (d) the branching structure of a tree. The human circulatory system is also fractal, as seen in Figure 3. These can all be understood as patterns that use minimal energy to achieve the most efficient flow through a system. Their self-similarity is self-evident: a part looks like larger parts, on all scales up to the whole. Mathematically, fractals are defined as self-similar structures on all scales, but for natural fractals there is always a limited range of self-similarity.

The venation pattern within a leaf looks rather like a river basin (Figure 4), and it is explained as a natural result of the flow of sap and the effect of auxins on the differentiation of water-conducting elements in plant leaves. The patterns shown are all robust constructions, generated by different processes of self-organization, each functioning effectively in achieving a particular result. They are examples of natural design in both living and non-living systems. Of course there are many other forms beside fractal structures that are robust and efficient. In plants there are the patterns of arrangement of leaves on the stem, called phyllotaxis, which can also be understood as energetically efficient in their production and functional in effectively catching photons for photosynthesis (cf. Douady and Couder 1999; Goodwin 1994, 2006). Animal form is equally complex, but the principles whereby structures such as somites, limbs and eyes are made are becoming clearer as a result of the union of previously separated disciplines, such as evolution and development. As we learn more about the dynamics of these processes, how they come into being, we realize that there are principles in operation in nature that produce forms characterized by elegance and integrity that are models for our own designs and constructions. This fascinating subject of natural design, how we humans can achieve the energy efficiency, the resource sustainability, and the functional beauty of natural phenomena, will be explored throughout this book as a major theme.

2. Health: Coherence with Meaning

Dynamic indicators of wholeness and health

In medical science, the rhythmic beating of the heart has been regarded as the essence of regularity and order in the human body in which the steady, reliable contraction of this remarkable muscle keeps the blood circulating to all parts of the body for seventy years or more. Cardiologists have been studying the pattern of heart contractions since the early years of the twentieth century by recording the electrical potential changes that occur over the body with every contraction of the heart. These electrocardiograms reveal many different patterns that are characteristic of a variety of heartbeat abnormalities, described as cardiac arrhythmias. Instead of the regular beating of the healthy heart, contractions can be clustered in various ways, such as in twos or threes; they can run faster or slower than normal; or there can be sudden, unexpected contractions that trigger an episode of fibrillation, a rapid and ineffective beating of the heart. Unless quickly corrected ('defibrillated') such episodes can be fatal, resulting in what is known as sudden cardiac arrest.

When deterministic chaos appeared on the scientific scene in the 1960s and it was realized that it is a typical pattern of behaviour for many dynamic systems, it wasn't long before people interested in heart dynamics began to ask if there were conditions of disease that could be characterized as the presence of chaos. Was sudden cardiac arrest, for instance, due to some disorganization of the heart rhythm that could be described as chaos? One observation that cardiologists had already made was that healthy children tend to have rather irregular heart beats, but this did not fit with the idea that chaos might be associated with disease. It pointed in the opposite direction, leading to the surprise that emerged in the 1980s.

The rule that the heart rate of a healthy individual is reliably regular when they are sitting or lying or walking is perfectly correct. However, what is being measured is the average or the mean heart rate. But if you look at a series of heartbeats for any one of these conditions, as recorded in an electrocardiogram, there is considerable variability in the interval between successive heartbeats. What came as something of a surprise was that this variability is significantly greater in healthy individuals than in people with various types of heart condition, such as cardiac arrhythmias or congestive heart disease. In the latter cases there is more regularity and order in the heart rate than in healthy persons. This is a case in which too much order, or the wrong kind of order, is a sign of danger!

It is possible that the irregularity of the interbeat intervals in healthy individuals is a kind of 'noise' resulting from the sum of influences exerted on the heart by other systems of the body — the nervous, respiratory, endocrine, muscular and other systems whose activities modulate heart rate. On the other hand, healthy variability might carry within it some signature of a subtle dynamic order that transcends the collective influences of these other parts of the organism. Peng *et al* (1996) claim that the variability of the interbeat interval does not have the characteristics of noise, but of deterministic chaos. The order manifested by chaos is indeed subtle, the dynamics being characterized by irregularity that is unpredictable but mathematically determined by the properties of strange attractors, the mathematical descriptors of chaotic dynamics that constrain the trajectories of motion within bounds as described in Chapter 1. The functional interpretation of this unexpected physiological behaviour is as follows. The healthy heart maintains continuous sensitivity to unpredictable demands on it from the rest of the body by continuously changing its rate so that it never gets stuck in a particular pattern of dynamic order. The presence of a chaotic element in the heart beat results in a type of long-range order in time that maintains a coherent balance of frequencies in heart rate. This is self-similarity in the time domain, complementary to the spatial self-similarity described in Chapter 1. A diseased heart, on the other hand, tends to fall into a pattern of order that fails to respond to the body's constantly changing needs. We thus get the notion of dynamic disease in which too much, or inappropriate, order is indicative of danger. Analytical methods that allow one to distinguish between different conditions of health and disease through the study of heart rate variability. This includes examination of changing patterns of heart rhythms as people get older, which indicates that there is a progressive loss of heart rate variability with age, the body losing its sensitivity and adaptability to unexpected demands as it ages (Costa *et al* 2002).

Do healthy people all share the same dynamic signature of health, or are they healthy in distinctive ways? This question was addressed by Ivanov *et al* (1996) in a study of people suffering from sleep apnoea (interrupted breathing during sleep) compared with matched healthy controls. They found that while each healthy individual has a distinct pattern of variability, they all share the same generic signature of subtle dynamic order that is characteristic of chaotic systems, characterized by self-similarity and the occurrence of a well-defined scaling law of variations. Individuals with sleep apnoea do not have this pattern. The property of healthy variability can be characterized as a type of long-range order or coherence that maintains a subtle balance of activity in the heart such that a series of short interbeat intervals tends to be followed by longer intervals, and vice-versa. This reflects the property of self-similarity or fractal structure that characterizes chaotic dynamics, as already described above.

The origin of chaotic behaviour is not yet clear. It appears to reflect a property of the whole organism that transcends the behaviour of its parts. This points to a holistic aspect of the organism with causal efficacy; that is, the observed dynamic is an emergent property of the whole that affects the parts, maintaining a condition of coherence throughout the organism. These studies are of considerable interest and importance in indicating ways of diagnosing different conditions of the body by a detailed dynamic analysis of particular physiological variables, which is now a very active area of research.

Seeing the condition of the whole in the properties of a part is the diagnostic procedure used in many healing traditions such as Traditional Chinese Medicine, reflexology, and homeopathy. Traditional Chinese practitioners using pulse diagnosis claim to observe the state of the various organ systems of the body and their dynamic balance or imbalance, pointing to appropriate treatments. The study of heart rate variability as an indicator of health or disease, and the extension of this to other continuous dynamic variables such as blood pressure and respiratory rhythms, may indicate a convergence of Western medical with other healing traditions in working with both parts and wholes in diagnosis and treatment. Furthermore, there is now research that uses heart rate variability to measure the efficacy of therapeutic treatments such as daoyin healing, in which there is no contact between the therapist and the patient, with positive indicators of effectiveness (Low 2003). This opens up new avenues for assessing directly the result of therapeutic treatments for which Western medical science has as yet no explanation.

Health: a strange attractor?

The intrinsic variability described above that has been identified in a healthy heart is also observed in other physiological processes such as the activity of the brain, the hormonal system, the immune system, the gut, and so on. It seems that healthy functioning of the body is characterized by a combination of order and chaos of the type that has been identified in many natural systems. It is conventional wisdom that the body heals itself, so that after wounding or disease the body returns to a state of sensitive dynamic functioning that we characterize as health or well-being.

There are other attractors for dynamical systems than the strange ones that characterize systems with deterministic chaos. The pendulum of a clock with a source of energy to keep it going expresses a pattern called a periodic attractor, and when the clock runs out of energy and comes to a stop, that state is called a point attractor. These and another doubly-periodic pattern called the toroidal attractor were regarded as the major descriptors of dynamical processes before the discovery of chaos. If we average out the fine details of the heart beat we get a periodic attractor describing the conventional view of the rhythmic activity of the heart. However, healthy bodies have a chaotic component to their physiological functions, as we have seen, so we can say that the dynamics of health includes the characteristics of a strange attractor. When bodies heal themselves they are returning to their natural dynamic state of wholeness and health, an attractor that combines order and chaos in a distinctive way. With this behaviour comes long-range order or coherence that maintains sensitivity and dynamic balance in the body's diverse physiological functions, all coordinated with one another. Disease, within this perspective, is when a physiological system has fallen into a state of too much order, a dynamic attractor that has lost it strangeness! Another way of putting this is to say that too much order is a sign of danger, which has the ring of commonsense. We know that if our lives are dictated by too much order and routine, we are likely to lose our adaptability and our creative edge. So it emerges that chaos is a source of adaptive novelty, of creativity. In a creative world it therefore becomes obvious that there must be chaos for new things to emerge. Chaos doesn't explain the particular types of order that arise in natural systems, any more than randomness explains the types of novel order that arise in different species during evolution. It is a necessary, not a sufficient, condition for coherent emergent properties, of which health is one.

Seeing cancer in context

Cancer is one of those words that conveys a sense of doom and deep foreboding, reminding us of our relative helplessness in the face of disease and death. While it is always true that confrontation with the imminence of death presents us with the deepest possible challenge to the meaning of our own lives, there are different ways of responding to reminders of our mortality. In our scientific culture we have set ourselves the task of overcoming disease, ageing and even death, so progress in medical science is naturally measured in terms of extensions to life expectancy and reduced mortality. Mortality is measured as the number of persons dying of a specific disease (say lung cancer) per 10,000 of a given population, adjusted for age. The goal of medical treatment is to reduce this. The most dramatic reductions in mortality related to improvements in cancer treatment have occurred in childhood cancers, testicular cancer, leukemia and lymphoma. However, these are all relatively rare. Of the common 'solid' cancers only breast cancer has shown a significant fall in mortality (from 1987 to date) as a result of better treatment. The other common cancers, namely lung, gastrointestinal, prostate, head and neck and ovary demonstrate depressingly static mortality trends. Some medical practitioners have taken the view that high dose chemotherapy and bone marrow rescue treatments now advocated by some medical oncologists might be looked upon as the death throes of the contemporary paradigm.

What is this paradigm, and what alternatives may lead us in a more fruitful direction of understanding and action? Basically, what we are seeing is a conflict between a view of cancer that emphasizes simple causes that make possible its control and management, in contrast to one that sees it as a disturbance in the complex network of relationships within cells and with the body that calls for more holistic understanding and intervention. The conflict can be seen as one between a management model of health on the one hand, and healthy living in communities on the other. The goal is to seek a reconciliation of these. To do this, we need to look in more detail at what we know about cancers.

Solid tumours are aggregates of body cells that have organized themselves into structures that are distinct from, though related to, the tissues and organs in which they arise. They can be benign or invasive, the latter being the dangerous category since cells from these tumours can travel through the body and take up residence at sites in other organs where they form secondary centres of growth. It is the continuous growth of these cancer cells that threaten the person with cancer. Here are cells that are us, with our genes and living in our body, that have somehow escaped from the normal control processes regulating the birth and death of cells that maintain our different organs in their normal states of action and interaction, resulting in the overall coherence that we call health. How and why do cells of our bodies do this?

The dominant view of how cancers arise in the body is focussed on the role of genes in determining the behaviour of normal and abnormal cells. In the past thirty years a great deal of research has gone into attempts to understand the process of cell division, wherein one cell becomes two cells of a particular type. The cells in the basal layer of our skin grow and divide continuously to provide a constant source of new skin cells to replace those we lose through wear and tear at the surface of the body. Similarly, cells in the gut, lungs, breast, prostate gland, liver and other organs grow and divide to replace cells that wear out, maintaining these in a steady state of healthy function, carrying our their roles in the overall economy of the body.

When cells grow and divide, they must reproduce all their components (membranes, mitochondria, vesicles, and nuclei with their chromosomes containing the genes) as well as the dynamic organization of these interacting structures that arise from the activities of various types of molecule that pass to and fro between them. Cell division involves not just precise copying of the genes but also their exact distribution to the two new cells so that each has the normal two copies of each chromosome. Anything that disturbs these processes can result in either change in the genes by mutation, or some other disorganization of the genes on the chromosome such as a change in the number of gene copies or their relative positions on the chromosomes. It was discovered in the 1980s that there is a special class of genes whose job it is to check for such mistakes and to arrest cell division until they are corrected, or to abort the process so the cells with errors die. These genes are called the 'guardians of the genome.' Any damage to them can result in cells with disturbed activity patterns, such as failure to respond to the normal signals that balance the rate of cell division with rate of cell death in the body. This may lead to the formation of a cluster of cells that is not integrated into the normal spatial organization of a tissue, which can form a solid tumour. The 'guardians' may therefore be regarded as the management team whose job is to monitor the intricate copying and delivery processes of cell division, making sure that all members of the team do their jobs properly.

These observations gave rise to the theory that the primary cause of cancer is mutations in this special category of guardian genes, also referred to as 'oncogenes' (that is, cancer-producing genes). It is this theory that has failed to provide medical science with a comprehensive and effective way of explaining and treating cancer. Detailed studies of gene activities in normal and malignant cells of the human gut failed to show any consistent change in the level of expression of these oncogenes in cancerous cells (Zhang *et al* 1997). Furthermore, many of the most powerful cancer-producing chemicals do not cause gene mutation, so damage to oncogenes is not the cause of cancer. And finally, the curves showing the pattern of growth of many types of cancer in humans often don't follow the shape predicted by the gene mutation theory (Rasnick 2000). As is usually the case in science, it is likely that the oncogene theory of cancer is not wrong so much as too narrow and limited in its perspective, requiring a different approach in order to make sense of its insights within a more comprehensive context.

Cancer as genetic imbalance

Some recent work (Rasnick and Duesberg 1999; Rasnick 2000, 2002) presents a new theoretical context for an old theory about cancer, with interesting implications for a shift in emphasis from treatment to prevention. This starts by showing why the current theory of genetic mutation in a few key 'oncogenes' as the primary cause of cancer is implausible, using a now widely accepted theory of metabolic control to demonstrate that mutations in a few primary control genes will either kill a cell or produce only a small disturbance in its behaviour (Fell 1997). Instead of all cancers suffering damage to a small set of key control genes so that they show similarities of genetic disturbance, all cancers in fact differ from each other in the types and quantities of molecules they produce, and are distinctly different from normal cells. This requires changes in hundreds or thousands of genes, not just a few.

It has been known for nearly one hundred years that cancer cells have a large imbalance in their genes compared with normal cells. The most familiar example of gene imbalance, technically known as aneuploidy, is Down's Syndrome, in which chromosome 21 has three copies in each cell of the body instead of the normal two. No genes are missing; it is the presence of extra copies of genes in the cells that is the problem. This genetic imbalance results in a disturbance of embryonic development so that tissues and organs are altered in their properties. However, the cells of a Down's Syndrome individual are not cancerous. It takes a more extensive imbalance of genes than an additional chromosome to produce cancer. Cancer cells often have many copies of some chromosomes, additional fragments of others, loss of whole chromosomes so that some have only one copy, resulting in imbalance in thousands of genes. How could this come about?

Mutations in any of the 'guardians of the genome' can result in a cell that has some aspect of its normal organization disturbed. Two genes in this category have been linked directly with hereditary breast and ovarian cancer. Cells with defects in these genes (labelled BRCA1 and BRCA2) have a high frequency of breakage and abnormal re-splicing of chromosomes, resulting in elevated levels of aneuploidy, so that there is indeed a role for mutation in cancer production. However, in a study of 211 women with breast cancer, only three had defects in BRCA1, so there is no simple correlation between defective activity in this gene and cancer. Furthermore, there are many other disturbances that can start the process of chromosome replication going wrong, such as chemicals from the environment or radiation or any form of stress. It doesn't have to be mutation.

Once genetic imbalance (aneuploidy) starts, it will tend to get worse: the imbalance in chromosomes will result in further disturbances to cell division, so that there is a positive feedback effect. However, this is counteracted by the fact that aneuploid cells aren't as healthy as normal cells and the body will tend to get rid of them. In a simple and elegant mathematical description of the development of aneuploidy, Rasnick (2000) has shown how this process fits data on the age distribution of different types of cancer in humans better than the mutation theory. Furthermore, he explains a number of other classical experimental observations on human cells grown in cell culture, such as the limited number of cell divisions that primary cultures can achieve (the Hayflick limit) and the rates of transformation of different types of human cell to the cancerous condition, among other results. He also proposes a very useful experimental test of chemicals for their capacity to disturb cell division and produce aneuploidy.

A cell that has gone through the process of genetic destabilization and has arrived at a fully cancerous state of continuous cell division and invasiveness has effectively developed into a new type of cell. It has a unique genetic organization and specific protein markers on its surface, but it shares with all other cancer cells the ability to propagate its kind in a particular environment, that of its host. It has effectively become a new species of cell, a stable genotype with a corresponding cellular phenotype. As Rasnick (2000) has pointed out, this process of destabilization, reorganization, and adaptation to a particular habitat is similar to the process that is involved in producing a new species of organism. So we may see cancer as an expression of the innate tendency of the living process to explore the possibilities open to it and to express successful new forms of life in available habitats. What we are seeing here is the emergence of a new cell attractor, a stable and coherent dynamic life cycle.

A somewhat different perspective that supports this general viewpoint comes from the recognition that cancer is not a disease of cell multiplication, but a disease of cell differentiation. The only normal cells in the body that divide are called stem cells, which are either undifferentiated, or partially differentiated. Differentiated cells, like muscle, nerve, skin, connective tissue, and so on, do not divide. Oncogenes tend to be regarded as genes that control cell multiplication, so that when they mutate the normal regulation of cell division rate is lost and malignancy occurs. However, it is more likely that oncogene mutation interferes with the stability of the differentiated state in cells, so that they become only partially differentiated and can then begin to divide (Harris 1985, 2003, 2004). The number of genes involved in maintaining cells in a differentiated state is large, so that in general many of these stabilizers need to be altered before the cell can begin to divide at a significant rate. The reasoning here is clearly similar to that involved in the aneuploidy hypothesis, and for the same reason: cell states are not the outcome of activity in a few key control genes, but arise from coordinated behaviour in self-organizing networks of elements in which patterns of relationship are primary, not hierarchical principles of control.

This new/old approach to cancer shifts the emphasis from cure to prevention. First, the recognition that every cancer is different means that it will be very difficult to target cancer cells specifically with new drugs. This has been known for many years, but the hope remained that there would be a primary, common molecular signature of cancer because of its attractiveness for drug-based cures. Chemotherapy is based on the use of toxic chemicals that kill all dividing cells, normal as well as cancerous, with the hope that the treatment will destroy the cancer before it kills the patient. Second, the emphasis on the diverse factors that disturb cell division means that the multiple chemicals that pollute our environment need to be screened for their capacity to produce aneuploidy. Most of these don't cause genes to mutate, but may well disturb chromosome replication and so disturb the genetic organization of a cell. And finally, the phenomenon of cancer remission, in which the individual gets rid of the cancer spontaneously, needs to be much more thoroughly explored. Remissions can occur after various types of stimulus to the whole body, such as change of diet, change of lifestyle, and many other non-specific influences (Challis and Stam 1990). However, such general or systemic disturbances to an individual, which include personal loss and grief, can also initiate cancer.

A new approach based on many different levels — the molecular, the cellular, tissue, whole body, person, community and environment — needs to be developed. The challenge is to provide a conceptual framework based on the rich network of relationships that characterize living processes so that we can make sense of the structure of the complex pattern of activities within cells and understand the dynamics of its behaviour. The management model of cell organization fails to do this. What is now emerging in biology is a model of life at the cellular level that is much closer to communities of agents with a wide range of creative responses to different circumstances than the current deterministic model of cell behaviour under the control of genes. This makes our own experience of living in community more relevant to understanding the life of a cell. The essence of living appears to be similar at whatever level we encounter it, a viewpoint that will be developed systematically in Chapter 4.

The healing power of relationships

If communities are important in the health of individuals, then there should be evidence that the relationships people have with one another are significant in their resistance to, and recovery from, disease. Detailed evidence for this is provided by the American doctor Dean Ornish in his book *Love and Survival* (1997). He provides copious evidence from a variety of detailed studies demonstrating that the most powerful factor in the incidence of illness, premature death from all causes and recovery from serious medical conditions in contemporary society is the pattern of relationships between people. For instance, 2,300 men who had survived a heart attack were interviewed to find out what kind of social interactions they had. Those classified as being socially isolated and having a high degree of stress had more than four times the risk of death compared with men who had low levels of both stress and social isolation, even when taking account of other factors such as genetics, smoking, diet, alcohol, exercise and weight (Huberman *et al* 1984). That is, the strongest predictor of death was social isolation and stress, irrespective of lifestyle. These men were all on beta-blockers, widely prescribed by cardiologists after heart attacks. However, social style was more important than the drug in extending life expectancy for members of the group studied.

Another study (Marmot *et al* 1975) that arrived at the same conclusion was carried out on 11,900 Japanese who lived in Japan compared with those who had migrated to Honolulu and San Francisco. The incidence of heart disease was observed to be lowest in Japan, intermediate in Honolulu, and highest in San Francisco. These differences were not explained by differences in diet, blood pressure, or cholesterol levels, and the incidence of smoking was in fact highest in Japan! The investigators sought an explanation for this. They classified the Japanese-Americans in San Francisco in terms of the degree to which they had retained a traditional style of Japanese culture. Those who had retained their social patterns of interaction, their family ties and community relations had an incidence of heart disease as low as those living in Japan. On the other hand, those with the most westernized lifestyle had a three- to five-fold increase in heart disease.

A cancer study revealed a similar result. A group of women with metastatic (invasive) breast cancer, all receiving conventional medical care such as chemotherapy, surgery, radiation, and medications, were randomly assigned to two groups. One of these met together for ninety minutes once a week for a year, the other did not. Those attending the gatherings were encouraged to attend regularly to express their feelings about the illness and its effect on their lives within a sympathetic and supportive context that allowed them to talk about their fears of disfigurement, of dying, of being abandoned by their friends or spouse, and other concerns. These groups were led by a psychiatrist or social worker together with a therapist who had breast cancer in remission. The instigator of the study, Dr David Spiegel, had undertaken the work to disprove the idea that psychosocial factors could prolong the life of women with breast cancer. When he analysed the results of his study five years after conducting it, his opinion was completely transformed. Those women who had the weekly support group lived on average twice as long as those without this opportunity to share their feelings with others. This and other studies are presented in Spiegel's book *Living Beyond Limit* (1993).

As a result of his own research as well as that of many others, Dean Ornish has come to some remarkable conclusions about health and medical treatment. Here is a quotation from *Love and Survival:*

> Love and intimacy are at the root of what makes us sick and what makes us well, what causes sadness and what brings happiness, what makes us suffer, and what leads to healing. If a new drug had the same impact, virtually every doctor in the country would be recommending it for their patients. It would be malpractice not to prescribe it — yet, with few exceptions, we doctors do not learn much about

the healing power of love, intimacy, and transformation in our medical training. Rather, these ideas are often ignored or even denigrated.

At the same time, Ornish is perfectly clear about the value of Western medicine in many circumstances, and is at pains to present a comprehensive, holistic context for the healing professions. 'Although I respect the ways and power of science, I also understand its limitations as well. What is most meaningful often cannot be measured. What is verifiable may not necessarily be what is most important,' he says in his book, and quotes a British scientist, Denis Burkitt: 'Not everything that counts can be counted.'

My own experience with surgery and medication is fully in accord with this recognition of the importance and value of the remarkable achievements of Western medicine. My life has depended directly on them at critical moments, and I continue to benefit from appropriate medication. The issues here are not either/or but both/and, keeping the whole spectrum of the living journey in view and seeking the most appropriate and effective action as it unfolds through myriad, diverse opportunities for learning and loving.

Healing with meaning

One of the many puzzling phenomena in the process of healing is what is known as the 'placebo effect.' This is characterized by people getting better in response to a stimulus that has no known physiological effect, such as an injection of a weak solution of salt (sodium chloride), a pill that contains nothing but sugar, or an action on the part of a healer that is not intended to heal but looks like a healing gesture to the subject. In seeking an explanation of this, it has been proposed that some people are simply highly impressionable so that they respond positively to suggestions about getting better. However, detailed studies of the placebo response indicate a much more complex picture of how and why people get healed. A detailed study of the placebo phenomenon and healing from a highly interesting perspective is presented in Daniel Moerman's *Meaning, Medicine, and the Placebo Effect* (2002).

In this book Moerman suggests that we think about placebo in a new way, illustrating this with an example of a revealing experiment. In this,

volunteers were asked to describe the effects they experienced as a result
of taking pills they were told contained either a tranquillizer or a stimu-
lant, though in fact the pills had no clinically active ingredients in them.
Moerman describes the test and its results as follows:

> A group of medical students was asked to participate in a
> study of two new drugs, one a tranquilizer and the other a
> stimulant. Each student was given a packet containing either
> one or two blue or red tablets; the tablets were inert. The
> students' responses to a questionnaire indicated that (1) the
> red tablets acted as stimulants while the blue ones acted as
> depressants, and (2) two tablets had more effect than one.

Moerman suggests that the responses by the students to the pills can
be explained by the 'meanings' in the experiment: (1) Red means 'up,'
'hot,' 'danger,' while blue means 'down,' 'cool,' 'quiet'; and (2) two
means more than one. These effects of colour and number have been
widely replicated. As Moerman observes:

> In the red versus blue pill study, we can correctly (if not
> very helpfully) classify the responses of the students as
> 'placebo effects' because they did indeed receive inert tab-
> lets; it seems clear, however, that they responded not to the
> pills but to their colours.

Moerman uses the term *meaning response* to define the physiological or psychological effects of meaning in the origins or treatment of
illness. We often interpret meaning as signifying, that is, discriminating
between alternatives, which is the rational or intelligible aspect of a
statement. However, meaning also has a feeling or affective aspect to it,
as when we ask, 'What does this mean to you?' in the sense of: 'What
does it matter to you?' We are then exploring how someone feels about a
particular issue, as in the question, How do you feel about taking a drug?
What does it mean to you to be receiving a drug? This sort of question
is tuning in to peoples' lifestyles, what makes them feel good and what
worries them, what they do and do not believe in: that is, their personal
and cultural belief systems.

Moerman elaborates further on this notion of the meaning response:

Anthropologists understand cultures as complex webs of meaning, rich skeins of connected understandings, metaphors, and signs. In so far as (1) meaning has biological consequence and (2) meanings vary across cultures, we can anticipate that biology will differ in different places, not because of genetics but because of these entangled ideas; we can anticipate what Margaret Lock (1993, 1998) has called 'local biologies.'

Whereas there certainly are biological universals across cultures, such as the presence of hormones, neurotransmitters and antibodies in all humans, there are also differences in which signals or stimuli influence the levels of these molecules in people from different cultures because of the meaning associated with these signals or stimuli.

Connecting physiology with feeling and meaning

How are we to explain the way in which an emotional or feeling response to something meaningful can affect our physiological state and so influence healing processes in the body? There is a direct connection between molecular networks in the body and our emotions. A familiar example of a type of molecule that affects feelings is the endorphins, those endogenously produced opiate-type molecules that give us a high, elevated or spaced-out sensation. These neuropeptide signalling molecules activate receptors in the nervous system, in the immune system, and in the endocrine system, all of which 'talk' to each other in a coordinated way, as shown by the studies carried out by Candace Pert and her colleagues during the 1980s and 90s (*Molecules of Emotion*, 1993). The result of these discoveries was the realization that different physiological systems, previously thought to be independent of each other and dedicated to separate activities in the body, are in fact all interlinked in a manner that achieves overall coordination of physiological behaviour. The new integrated perspective that resulted was called psychoneuroimmunology, to which can be added the endocrine or hormonal system. Here is a foundation for putting material activities or molecular processes together again with feelings, after some four centuries of separation in western science.

The way this coordination is achieved is through the usual processes of feedback, whereby specific molecular signals increase or decrease the activities of their target organs or cells. Monocytes are roving immune cells that digest foreign bodies such as invading bacteria or alien proteins. Their activity is influenced by neuropeptides that are produced in response to emotional states. A person suffering from chronic stress can produce specific neuropeptide signals and hormones that depress the activity of these immune cells, making the person vulnerable to disease. This type of coordination between different physiological activities and emotional states in an individual results in the union of body and mind so that it becomes clear that the experience of fear, anxiety, anger, joy, or beauty can have immunological correlates. Furthermore, we can see how the experience of trauma, either psychological such as sexual abuse, or physical, as from a serious accident, can be recorded as a long-term change in the tuning of the psychosomatic network of signalling molecules or in the structural aspects of the body, such as collagen in connective tissue. There is no aspect of the body that is immune to such influences. In consequence, we need now to recognize that what we call the mind is an embodied process of continuous modulation of the body that records the life history of the individual. Most of this embodied knowledge is largely unavailable to conscious inquiry, so that we are unable to describe precisely how it is that we can do things like ride a bicycle or play a guitar or see the funny side of a story. Even more so are we unable to describe why we may be suffering from anorexia or from intermittent migraine headaches. These conditions are telling us something about our embodied experience, but it can be very difficult to get at the deeper meaning expressed by the body and so to take appropriate action to achieve coherence and well-being. Nevertheless, the perspective of embodied mind and deep knowledge in our bodies as intelligent agents opens up the possibility of a creative dialogue of participation with our lives rather than the repressive attitude of control.

Food and health

A learning journey that we have been on recently in the industrialized West concerns the issue of food and health. This is not simply about our own personal health, but the health of the farms on which food is produced, the health of the countryside and the health of the communities of farmers who produce it. Deep unease and concern about the way we handle and feed farm animals, and manipulate the food we eat, has emerged in the UK, Europe, and now North America, particularly with the outbreak of BSE (mad cow disease), but also with various forms of bacterial contamination. The tradition of reductionism in Western science led to the belief that it makes no difference what proteins you feed to animals, since these are broken down into their constituent amino acids during digestion, before being used in the body. BSE showed that this assumption was tragically incorrect. Some proteins are stable, are not digested, and can be taken up by the animal, one of which is the agent that causes mad cow disease. This might have been taken as a warning not to make simplifying assumptions about complex processes when food and health are involved. However, there is another experiment going on that is based on the belief that it is perfectly sensible and safe to modify the genetic inheritance of organisms by design and manipulation in order to satisfy human needs. In agriculture this has given rise to the emergence of biotechnology companies that are redesigning crop plants for a variety of purposes.

The traditional view of heredity that derived from Mendel's classic studies of patterns of inheritance in sweet peas saw genes as stable units of inheritance that are transmitted from generation to generation, performing a well-defined and constant hereditary function irrespective of the organism in which they are found. The presence of particular genes were seen to be responsible for generating particular characters, conventional examples being genes for flower colour, height of plant, number and size of seeds, and so on.

This view underwent a profound transformation in the 1960s and 70s when it was realized that genes are much more mobile and sensitive to context than had been previously believed. 'Jumping genes' or transposons were discovered, first by Barbara McClintock's pioneering (and initially disbelieved; see Keller 1983) studies on maize and then by many others. These mobile genetic elements move around the chromosomes and have different effects depending on their neighbouring genes. It then emerged that bacteria exchange genes via viruses and virus-like elements called plasmids. This has come to be known as horizontal gene transfer since it is not the traditional passage of genes from one generation to the next but movement of genes between members of the same generation. It was realized that the genotype of an organism is much more fluid and changeable than had previously been believed, and that gene products are variable and sensitive to context (Ho 1998, 2003).

This means that the genotype does not determine in any fixed and pre-specified way the set of molecules produced in a cell that contribute to its characteristics. There are many more phenotypic possibilities open to a cell and an organism than are defined by one specific way of reading the base pairs in its chromosomes. Despite these revelations, the bio-technology industry continues to say that the construction of transgenic plants is an exact science in the sense that specific genes with specific, known consequences can be transferred from one species to another to produce predictable properties in crop plants.

Paradoxically, it was precisely the discovery of the fluidity of the genomes of bacteria and viruses in the 1970s that made biotechnol-ogy possible, allowing the design of vectors (plasmid-like constructs) to carry genes from one species to another. However, biotechnology extended the natural processes by carrying out gene transfer between completely unrelated species, such as transferring the gene for resistance to freezing of tissues in the Arctic flounder to strawberries, to confer frost resistance. The vectors also carried genes from viruses combined with those from bacteria into the genome of crop plants such as soya, oil seed rape (canola) or maize, in order to make them resistant to pesti-cides or insects. These combinations are extremely unlikely in nature, so genomes were being mixed in totally novel combinations. Furthermore, the insertion process is random so that the position of the transferred genes in the plant's chromosome is unpredictable. The predictable result is genetic instability.

Owning life through patents

Biotechnology thus opened a Pandora's box of unknown possibilities and potential problems regarding the properties of genetically engi-neered crop plants and their interactions with traditional crop species, with weeds, and with insects — that is to say, with the other members of the ecosystem in which they are grown. Transgenic crops often have unexpected instabilities, so that the character that should be expressed is not, the crop fails in some unpredicted manner, or unanticipated properties are expressed. There is also the danger of gene transfer from genetically modified to GM-free crops so that contamination occurs, as has recently been observed in Mexico for native land-races of maize, now contaminated by genes from transgenic maize with insecticide genes inserted (Quist and Chapela 2001). However, the origin of the contaminating genes, whether from Monsanto or another biotech company, cannot be identified because the precise gene sequence used in the vector is a trade secret! This is the corporatization of science so that knowledge is no longer shared freely, as in traditional agricultural communities and in scientific research. It is now used for profit through a process of knowledge control (patenting of life forms) in which actual scientific knowledge is not only suppressed but distorted, as in the biotech rhetoric that its procedures are exact and predictable. In fact, it is now widely recognized by scientists that organisms are intrinsically unpredictable in their responses to any manipulation, for the reasons described previously relating to the complex dynamics of living processes. Despite these dangers, there have been no systematic, officially sanctioned and published studies of the effects of GM products on human or animal health, and no legal liabilities on biotechnology companies in case of damage to human health, genetic contamination of other crops, or for failure of genetically modified plants to produce the results claimed for them.

We have come to know well the properties of the food we have been producing and eating for centuries. Food science has made it possible to examine in quantitative detail the composition of different foods and to improve the means of storing it so as to prevent deterioration or bacterial and fungal infection, which can make food dangerous to eat. We know what methods of food production and treatment are safe, providing us with a variety of wholesome sources of nutrition. There is enough productive capacity to feed humans worldwide. As stated by James Wolfensohn, President of the World Bank: 'Even though global food output is adequate to feed the entire world's population, 800 million people are going hungry because they cannot afford to buy the food they or their families need.' Hunger and starvation arise because of inadequacies in food distribution and because of poverty, not because of food scarcity. Furthermore, farmers have learned various ways of working with the land so that soil quality is maintained and farms remain an integral part of the broader ecosystem of forests, rivers, wetlands and grasslands, preserving the diversity of species required for ecological health.

So why do we need biotechnology? It is of no value to consumers, and simply presents them with unknown risks and dangers. Consumers have expressed a desire to have GM products labeled so that they can make a choice. The argument from the biotech companies is that their transgenic crops are both more productive and more ecological than traditional crop varieties, conferring an advantage on farmers and on nature through reduced herbicide use. They claim also that consumer suspicion is based on ignorance of the benefits their products will bring to agriculture. However, after some ten years of this experiment in food production, the evidence is otherwise. Whereas it is true that some farmers have benefited from the use of GMOs because of higher yield and the use of less labour in growing the crop (thus reducing the number of people employed), the overall experience of farmers is that genetically modified crops give poorer harvests than conventional varieties, yet the seed costs more. As stated by Ann Clark, Professor of Plant Agriculture at the University of Guelph in Canada: 'GM crops are not benefitting farmers, the intended beneficiaries, in yield, reduced pesticide or herbicide use, or profit. The share of gross income per acre required simply to buy seed plus chemical has risen by more than 50% between 1975 and 1997, from 9.5 to 16.9% for maize.'

The same result holds for GM cotton in India, where the supposedly insect-resistant crops have failed dramatically to perform as claimed and many farmers have suffered substantial losses (Shiva 1993, 2002). Herbicide use has not decreased overall, due to a number of factors including the development of pesticide resistance in weeds so that more or different herbicides are needed to control these 'superweeds.' Given all this evidence that GM crops fail to produce any overall benefit to farmers, the public, or the environment, one wonders why the biotech companies are so persistent in pushing their products.

The basic reason for this is the potential financial profits that can be made by any company that succeeds in controlling substantial amounts of the world food commodity market. We consume a great deal more food than we do drugs, so the food market is worth very much more than that for pharmaceutical products, whose profits are vast. The basic control factor for drugs lies in patents, which give companies twenty years of exclusive right to marketing their products. Control of food commodities requires similar ownership of crops produced. However, farmers have always retained seed from one harvest to plant the next. Furthermore, as a result of centuries of selection of seed from plants that perform particularly well under specific environmental conditions, a great diversity of crop varieties has been produced by farmers using traditional breeding techniques of cross-pollination and selection.

Sustainable and unsustainable agriculture

Industrial agriculture altered these traditional practices in two ways. First, instead of planting a variety of crops to guard against unpredictable weather conditions and insect attacks on a particular crop, farmers were encouraged to plant the same crop over enormous fields, protecting it with herbicides and insecticides, watering from groundwater sources, and using heavy machinery for preparation of the soil, weeding and harvesting. Second, fertilizers were used to boost production, which also required more water. Fertilizers, pesticides and herbicides altered the quality of the soil and also contaminated rivers and groundwater. They also are the cause of much disease in humans who are exposed to them simply as a result of living near farmland treated with pesticides and herbicides. The evidence is that they can damage the nervous system, can cause cancer, and disrupt hormonal systems, but there is no adequate protection of people from them in the UK, as revealed by Georgina Downs, the activist who suffered for years from their effects (Leake 2006).

Despite these radical changes in farming practice, farmers could still use seed from one crop to sow the next. The exception to this was hybrid varieties that did not breed true so that the farmer had to buy seed from companies every year. When genetic modification became possible it was realized that the advantage of the hybrid to the seed company could be extended to all genetically engineered plants through ownership of the plant variety by patent. Biotechnology is industrialized agriculture plus ownership. The biotech company not only owns the seeds produced by a transgenic crop, so the farmer has to pay even if he harvests seeds from his own crop; but at the same time the company can sell him its own herbicides, to which its genetically engineered plants are resistant, as in the case of Monsanto's Roundup Ready soyabeans.

A remarkable legal judgment has emerged in Canada regarding this form of ownership. Percy Schmeiser is a farmer who has habitually saved seed from his own crops. He has never planted GM crops on his land, but his neighbours did. As a result, Schmeiser's crop of canola became contaminated by GM pollen from neighbouring farms. Monsanto claimed ownership of his seeds because of the presence of 'their' transgenes and took Schmeiser to court. The judge ruled that intellectual property rights on genes is a higher form of ownership than Schmeiser's land and crop rights, so the ruling was that he should pay Monsanto for their contamination of his crop, which incidentally made it worthless for the GM-free market. This is patent law gone quite mad, a kind of Alice-in-Wonderland inversion of good sense. The case then went to the Canadian Supreme Court, which ruled 5–4 with a rather ambiguous judgment: Monsanto's claims to ownership of the canola crop was upheld, but Schmeiser was not required to pay any fine. This indicates the level of uncertainty that prevails in this area, which continues to attract controversy.

Monsanto urges farmers growing their GM crops to keep an eye on their neighbours to see if they are infringing copyright by using GM seed that they have not paid for, and to inform Monsanto if this is the case, for which they receive a financial reward. This policy of encouraging informers is another consequence of extended ownership, undermining the traditional cooperative solidarity of farming communities in the most pernicious way. Systems of political informers used by repressive regimes have always been condemned by democratic governments, but it seems that the solidarity between the corporate sector and contemporary 'democracies' is such that these abuses of civil rights and their damaging influence on community relations are sanctioned in the name of technological progress.

As stated, GM crops perpetuate the present system of industrial agriculture. This is highly fossil-fuel dependent for production and distribution of fertilizers, (mainly phosphorus and nitrogen), herbicides, and pesticides, for the machinery required to run vast monoculture farming systems, and for transportation, handling, and packaging of most foodstuffs. Whereas it is possible for the average family of four to save around 8,000 kWh per year by insulating their house and to save some 6,000 kWh of fuel costs by getting a more efficient car, they can save around 30,000 kWh per year by buying locally produced organic food! (Berkes *et al* 2003) This is because of the intensely energy-wasteful nature of modern food production and distribution. Clearly our food production system is the least sustainable aspect of current industrialized lifestyles. And there is no way in which biotechnology could possibly alter this. Since the goal has to be sustainable agriculture, we can dispense with the whole issue of transgenic crops as anything other than a very minor aspect of food production systems of the future. Fossil fuel prices will start rising dramatically in the immediate future, as we pass through the peak of oil production and experience progressive shortages, making industrial agriculture uneconomic as well as unsustainable.

Healthy ecosystems

It now becomes clear that genetic modification of crop plants poses a spectrum of dangers that include not only the immediate effects on quality of food and farming ecosystems, but also the communities of farmers in whose hands lies the security of food production throughout the world. If ownership of the seed for major food commodities falls into the hands of a few transnational companies, we will no longer have any possibility of realizing a diversified and ecologically sustainable agriculture, which is the only long-term, locally secure solution to food production. Just to illustrate the contrast between these different scenarios, let me describe two methods of production of one of the staple food commodity crops in Africa, namely maize.

Maize has always been threatened by infestation by an insect, the cornborer. The biotech solution is: use genetically modified maize with a bacterial gene that produces an insecticide in all the cells of the plant. This is the gene that failed to protect cotton in India, and has variable effectiveness. Furthermore, since the insecticide is always present in all the tissues of the plant, but in variable concentration, insects eating the plant are exposed to different doses of the poison, often in sub-lethal amounts. This accelerates the development of genetic resistance to the insecticide in the cornborer, so in the long term the method fails. Furthermore farmers use the same variety of genetically modified plant irrespective of the local conditions, and continue with fertilizer treatment and irrigation on which the biotech variety is dependent.

Compare this with an alternative procedure, which was developed in an ecological research station in Africa. The maize is sown along with a legume, desmodium, which grows between the rows of maize as a ground-cover crop. It is a legume that fixes nitrogen from the atmosphere and leaves it as fertilizing nitrate in the soil. Desmodium produces a volatile material whose smell repels the cornborer. Around the edge of the field the farmer grows Napier grass, which attracts the cornborer as it is its natural host. The cornborers lay their eggs in the Napier grass, which has adapted to infestation by producing a sticky secretion which prevents a significant fraction of the cornborer larvae from developing, allowing sufficient numbers to survive so that the species persists. The farmer can grow whatever variety of maize does best under local conditions, gets a good maize harvest with soil that has been fertilized for the next crop, and can also harvest the Napier grass as fodder for the cattle. This so-called push-pull method is ecologically sustainable, allows farmers to plant locally adapted varieties of maize, and preserves the farmer's right to plant his own seed, on which his security and independence depend.

The biotechnology industry seeks to make massive profits in the short term through ownership of genetically modified seed and control of the crops farmers grow, while encouraging the unsustainable practices of industrial farming. There may be some transgenic crops that really do improve yields for farmers and help protect the soil. However, these are not the ones currently pushed by biotech companies. And ecological solutions of the push-pull variety are of no interest to the biotech industry because they cannot be patented. Clearly there is a deep conflict between the interests of those promoting genetic engineering in its present form, and the long-term health and vitality of crops, ecosystems, farmers, consumers and communities.

The important issues connected with biotechnology are being addressed by scientists who are blowing the whistle on these practices, and also by public action which is becoming much more discriminating in the acceptance of science applied to the complex systems on which the quality of life depends, in this case the quality of the food we eat. Civil disobedience has always been a significant social factor in correcting deficient democratic procedures. This has re-emerged in the face of what appears to be collusion between government and the biotech industry that threatens human, community and ecological health. What people are expressing here is dissatisfaction and suspicion of the basis on which judgments by many politicians, scientists and corporate leaders make their decisions. This unease goes much deeper than the scientific and technical details concerning innovative technology. It involves issues with which science is failing to come to terms that are particularly relevant to the crises we currently face. This has to do with ethical issues and the distinction between quantity and quality, in particular the inadequacy of quantitative measurements to evaluate the quality of life on which health depends. There are also deep issues concerning how people experience meaning in their lives through complex relationships with others and with nature. Is there a more comprehensive way in which we can gain relevant knowledge for engaging in responsible forms of communal action, participating cooperatively with the natural world to achieve healthy and sustainable lifestyles? This is the question explored in the next chapter.

3. Science with Qualities

Separating quantities from qualities

It was assumed at the inception of Western science in the sixteenth century, with the systematic studies of Galileo on cylinders rolling down inclined planes, falling and projected bodies, and the motion of the moons of Jupiter, that the only reliable knowledge we can have of the natural world is of measurable quantities such as mass, position, velocity, acceleration, and so on. The role of science was to discover the logical, mathematical relationships between these variables that underlie regular behaviour in nature. Qualities such as experience of colour, of odour, or a sense of beauty at the spectacle of the planets or the elegance of the equations that describe their motion are 'subjective' and cannot be used as reliable indicators of natural process. The distinction was later made by the philosopher John Locke between what he called 'primary qualities' of nature, which are measurable and objective, and 'secondary qualities,' which cannot be measured and are subjective. Scientific study is restricted to a systematic study of the former. This was a very sensible and useful assumption to make in order to examine one aspect of nature, its orderliness and predictability, in a systematic and logical way.

However, it is clear that this distinction presents us with a problem which at first was not very serious but which has become more and more acute as Western science has proceeded to extend its domain of inquiry to more and more aspects of nature. We live our human lives primarily in terms of qualities, not quantities. Our relations with each other depend deeply on our experience of one another, whether we have relationships of trust or suspicion, whether we believe that someone is truthful or honorable or is unreliable. We extend this type of qualitative judgment to our relationships with domestic animals such as sheepdogs or horses or oxen. People who depend on the natural world for their livelihoods make very extensive use of 'subjective' judgements in deciding on, say, the quality of soil for producing different crops or the quality and balance of an ecosystem to determine which and how many animals to hunt or wild plants to harvest. There are quantitative aspects to these evaluations, but no sharp separation between quantity and quality in assessing the properties of the complex systems on which our lives depend.

From control to participation

Chaos and complexity theory make it clear why this must be the case. We now know that we cannot use quantities and mathematical relationships to predict the behaviour of complex systems, both because they are likely to have the property of sensitivity to initial conditions and because the behaviour of the whole may be an unpredictable emergent property of the relationships between its parts. These properties are revealed by their qualities as much as by their quantities. A farmer assesses the quality of his soil in terms such as 'friability,' its texture and feel in the hand. At market, purchases are made not simply on the basis of the age and weight of an animal but in terms of its 'conformation' and the quality of its movement and behaviour. In the context of medical treatment and therapy, the health of an individual is assessed not simply in terms of temperature, blood pressure and blood composition (including cell counts of different type) but in terms of complexion, posture and tone of voice. These are indicators of the condition of the whole being. Quantities and qualities together are used to assess relevant properties of people, of other animals, of landscapes and of ecosystems. These are all complex systems which we now understand are not controllable or predictable or manipulable, except in limited degrees. But our lives depend on them. So how are we to relate to them? Traditional wisdom, and now new insights in science, tell us that, since we cannot control them, we must learn to participate in their processes using all our skills of quantitative and qualitative observation to assess the effects of our actions and the impact we are making on them. Instead of assuming that scientific knowledge allows us to get ever more control of nature, we must now accept that good quality of life depends upon relating to complex natural systems with sensitivity and attentiveness, and cultivating skills that conventional science has tended to demote to the realm of merely subjective judgement. What is now needed is an extension of science to include qualities.

A science of qualities

The term 'a science of qualities' sounds like a contradiction to Western scientific ears, since qualities were banished from science four centuries ago. However all scientific assumptions are tentative and should not be allowed to outlive their usefulness. We are now witnessing the consequences of ignoring qualities in science in the loss of habitat, species, health and quality of life generally, as described in connection with industrial farming and biotechnology in the previous chapter. A strictly quantitative approach to nature has given us the ability to produce vast quantities of consumer goods and wealth, but has resulted in the destruction of the quality of life of more and more species and people the world over. How can we use the insights of science itself to transform this situation dialectically? We can examine the validity of the assumption about qualities and see if it is necessary for scientific knowledge of the world.

The essence of scientific procedure is the exploration of the world using methods that have been agreed by a group of people who constitute a community of inquirers. As noted earlier, in this community there are no authorities; anyone can challenge accepted theories, hypotheses, models, assumptions and procedures, and introduce others that produce consistent results and make sense of observations. If someone challenges a deeply-held conviction, it is to be expected that there will be a lot of resistance to change. Thomas Kuhn (1960) explored this in connection with what he called normal versus revolutionary episodes in science, times of resistance and periods of change. He pointed out that for change to occur, it was necessary for there to be crucial anomalies between accepted theories and observation, and that an alternative interpretation of the phenomena (a new 'paradigm') must be proposed that allowed scientists to examine them from a different perspective. How does the current situation in science appear with respect to a change of assumption about the status of qualities in relation to the study of natural phenomena? There is no doubt that this is a big one, so the alternative to procedures using only quantities and their relationships must be pretty convincing, and urgent.

There are many examples of current scientific studies that attempt to bring qualities into the accepted procedures of science. Qualitative research is found in many studies of healthcare that transcend the conventional 'medical model' of disease. These allow personal reporting on well-being as evidence for the efficacy of complementary therapies, not just the usual double-blind trials based on quantitative measures of response to treatment. However, there is still considerable resistance to these procedures on the grounds that there is no test of consistency between the reporting of different individuals about their state of health. Scientific procedures seek evidence that the evaluations of different individuals about a particular quality of experience are consistent. This evidence is now available from research on the capacity of people to assess consistently and consensually observed qualities.

Extending 'objectivity' to qualities

In conventional science, what is called 'objectivity' is actually consensus between different subjects on a measured property of some natural process, such as the position or the velocity of a planet during its motion. Different subjects reach agreement about such properties by using previously agreed methods of observation and measurement. This process involves consensus, which allows for disagreement but seeks accord. A group in Scotland carrying out research into the capacity of people to reliably assess the quality of experience of farm animals has now shown how consensus can be assessed among people who are evaluating qualities.

The procedure developed was based on methods of evaluating the quality of food and drink, which is routinely used. A group of eighteen people, with no particular experience with farm animals, was shown a series of videos of twenty individual pigs behaving under standard conditions: a pen with straw, familiar to the pigs, into which a person enters and the pig can then relate to this person in any way it chooses. The people watching the video are asked to write down, independently of each other, terms which they think describe the quality of experience expressed by the pigs. These are freely chosen by the viewers and include terms such as nervous, laidback, pushy, aggressive, playful, cool, and so on — whatever comes into the viewer's mind as an appropriate descriptor of the pig's behaviour and expresses its subjective experience, as we would use for other humans. Each person then lists all the terms they have used for all the pigs, with a line of the same length beside each term, and copies this list so that there is one for each pig. The videos are then shown again and the observers, again independently of each other, put a tick on the line beside each descriptor to indicate how much of this quality they think describes the experience of the pig during the interaction with the person in the pen. This data is then used to carry out a statistical evaluation of the extent to which observers agree on the quality of experience expressed by each pig. The analysis is designed to look for clustering of the points in the multidimensional space which describes the results of the evaluations, and to compare this with the frequency of clustering that would occur with randomly assigned points for the pigs.

The results of this study (Wemelsfelder *et al* 2000) are quite striking. There is a high level of clustering for the experimental data, showing that there is a large degree of consensus between different individuals in their evaluations. Furthermore, the different terms used by different individuals tend to form coherent descriptors of the pigs, with words such as nervous, tense, and withdrawn tending to fall near one another on a principle coordinate display of the results, while terms such as pushy, aggressive and boisterous are located toward another pole of this semantic space. This shows that there is consistency of judgement regarding quality of experience observed in this experiment. Wemelsfelder (2001) argues that this provides evidence that observers are seeing a property of the pigs themselves, rather than simply projecting their own feelings on to the pigs, just as the subjective experience of weight is taken as a property of the object experienced, not simply an idiosyncratic projection of the observer on to the object. Furthermore, the observed quality of experience of the pig relates to the whole animal and the way it relates to its world. This capacity we have of direct knowing about the qualities of the whole is often described as intuition, non-inferential perception. A basis is thus provided for extending the procedure of scientific consensus of observed properties of natural processes from analysis of quantities, relating to the properties of parts, to the intuition of qualities, which refer to the properties of the whole. Furthermore, these studies show that the use of qualities to assess differences between individual pigs is more discriminating than the use of quantities to describe behaviour (number of steps forward, backward, frequency of grunts, and so on). Qualities capture essential aspects of the whole expressed in behaviour that quantities cannot describe.

In one sense these results are not news to anyone who works with animals or with nature. Only scientists and philosophers are likely to challenge the implications of these conclusions, and for good reason. They will require considerably more evidence that the procedure for reaching consensus on qualities provides a sound foundation for including in scientific judgement about natural processes what has for so long been regarded as the purely subjective domain of human feelings. However, the consequences of failing to include qualitative evaluations into the assessment of the changes we are making to the natural environment and to cultures are ever more pressing and dangerous. We are experiencing global crises in health, in community relations, in agriculture, in habitat, species and cultural destruction, and in climate change, all as a result of the way we have chosen to see these aspects of culture and nature from a scientific and technological perspective. The separations that we have made between nature and culture, between quantity and quality, and between control and participation, are now causing serious problems. This raises deep issues about the way we are educating ourselves, the assumptions we are making about ourselves and our place in the world. How are we to transcend these separations and enter into a more participatory relationship with the other members of our planetary society? This will be explored in Chapters 5 and 6. But now we need to look more closely at questions regarding the origin and nature of qualities from a evolutionary perspective.

Where do qualities come from?

How can we account for the evolutionary origins of feelings and the experience of qualities in humans? Darwin considered that emotions arose during evolution and were selected because they confer an adaptive advantage on those organisms that have them. Fear adds to the capacity to escape from danger, while maternal love enhances the quality of care provided to growing and developing young. However, we saw in Chapter 1 that the possible adaptive advantage of a property does not account for its evolutionary origins. It is necessary to provide an account of how the property in question is possible, how it can arise within a particular type of organism or system. In all examples of emergent properties that we have considered, the phenomenon that we seek to understand can be seen to arise from related properties of the components of the system. In the model of an ant colony described in Chapter 1, individual ants have intervals of activity and inactivity. This pattern is not periodic, but it does involve movement and rest. It is not difficult to accept that particular conditions of interaction can produce periodic behaviour of the colony. Although unexpected, this does not involve something coming from nothing. The model suggests how there is consistency between levels of process, the patterns of movement in individual ants and that which emerges in the colony, despite an inability to predict the phenomenon. Science can cope with this kind of novelty, can make sense of it by demonstrating consistency between levels of action in the colony, even if there is no causal closure that allows the phenomenon to be predicted. However, if some property arises that cannot be understood in terms of constituent processes of similar quality, like the hardness of teeth and the hardness of calcium phosphate crystals of which they are made, or periodic movement in an ant colony from chaotic movement of individuals, then we get something like a miracle, at which scientific understanding boggles.

Now consider the case of feelings. It is often assumed that feelings emerge as an aspect of the experience of organisms with sufficiently complex nervous systems. But the components of the nervous system, neurons, are themselves made up of lipids and proteins and ions which, organized structurally and dynamically in a particular way, result in action potentials and propagating electrical currents. None of these components of the nervous system could be said to have any property that relates to subjective feeling, to personal experience of process. So to assume that feelings emerge from such complex systems implies that suddenly, at some point in evolution, a totally new property emerges that cannot be understood in terms of some similar quality that belongs to the components of the system from which the property emerges.

This dilemma has led many scientists and philosophers to take the view that emotions and feelings in organisms are not real in the sense that claws and teeth and hearts are real. Feelings are illusory, epiphenomena invented by natural selection because they aid survival. However, scientifically this is really a cop-out, an abandonment of the task of scientific explanation, just as the assumption that natural selection explains how emergent properties of organisms arise fails to recognize the need to provide a scientific explanation of how these properties are possible under the circumstances from which they arise. The study of emergent properties in complex systems forces the issue about the origin of experience of qualities that is an undeniable aspect of evolution, unless we deny the reality of our own experience. This is a price that most people are unwilling to pay, quite rightly; for denying what is our everyday experience is to be deeply alienated from the reality of our lives. The alternative to denying the causal reality of feelings and emotions is to seek to explain them as real emergent properties of organisms during their evolution. There is now a very active debate among scientists and others interested in the evolution of human consciousness concerning this issue of the origins and status of qualities of experience (cf. Griffin 1998, Silberstein 1998). Many others are keenly interested in ways of making sense of our experience as thinking and feeling beings. Among the recent developments in this field are contributions from workers in robotics and artificial intelligence/artificial life and developments of these ideas. These are clarifying why it is that distinctive feelings are associated with experience of qualities in terms of the specific forms of activity that are involved in generating these experiences. They provide some indication of how the distinctive behaviour of organisms in making sense of their environments could be involved in the generation of feelings.

The primacy of movement and intention in perception

There has been a long tradition in Western science that has regarded sense perception as essentially a passive recording of properties of the outside world on the specialized sites of reception in the organism. Vision, for example, has been understood as a process similar to the action of a camera in recording a visual image. The external scene enters the eye via the lens and the retina and is transmitted faithfully to the visual cortex of the brain, where it is recorded as an internal representation of aspects of the outside world in neural activity (cf. Lindberg 1976).

However, this approach to perception has been challenged by those who have recognized that organisms are agents that actively engage with their environments during sense perception; they are not passive receivers of sense impressions (see, for instance, Gibson 1979). Recently this perspective has been significantly developed in connection with what is called 'skill theory' (Clark 2000). Clark argues that different modes of sensory perception such as seeing, hearing, touching, and so on, involve specific skills whereby the organism engages with its environment in particular ways. For example, in seeing colour the organism scans an object using particular eye movements and also changes its relation with the object in order to test aspects of invariance in the experience. The colour red has particular properties that change as the incident light on the object, from the sky (blueish) or from an artificial light (yellowish), is altered during examination of the object by a subject. Furthermore, colour experience does not change when the subject covers her ears, but does when she blinks. And the particular distribution of red-sensitive cells in the retina has an effect on the experience of red as she moves her eyes. In attending to a sound, an organism uses different modes of discrimination, such as turning the head to identify the direction from which it comes or paying attention to the changes of pitch (frequency) in the sound, as in birdsong.

The distinctive forms of action in which an organism engages while actively exploring different sensory stimuli provide the basis, according to Clark whereby the modality experienced, whether colour or sound or touch, can be identified. He describes this as 'unmediated or non-inferential access to modality,' which he regards as the basis of phenomenal experience. 'The fact that she has direct unmediated access to certain distinctive physical or functional features of the visual encoding will force upon the agent the idea that there is a special sensational quality present' (Clark 2000). Thus it is the active participation of the organism in experiencing the world in different ways that gives rise to direct knowing that is the distinctive quality of subjective experience, what it feels like to see or hear or touch. Feelings in organisms arise from the exercise of particular skills in exploring the world. This suggests how to investigate the activities that are associated with different feelings.

Myin and O'Regan (2002) have extended and transformed Clark's ideas in interesting and suggestive ways. In doing so, they have made important connections with the philosophical tradition of phenomenology presented by, for example, Heidegger (1958), Husserl (1960), and Merleau-Ponty (1945/1976). Myin and O'Regan point out that skill theory presents a specific break from traditional cognitive science, which emphasizes neurophysiological processes as the basis of consciousness, including sensory awareness and associated feelings. However, there is nothing in these neural activities that allows us to identify the distinctive qualities of different types of experience, since they are all basically the same, reducing to electrical changes and ion fluxes in neurons and other types of brain cell. This is where neural reductionism fails to provide any basis for understanding and explaining basic aspects of the reality we know. Skill theory puts the emphasis in perception and experience on engagement with the outer world rather than on inner experience associated with neural activity. In so doing, it provides a basis for understanding how different modalities of experience are generated by organisms as active agents in exploring the world, and why we see the world 'out there' and not inside us, where the representations are supposed to be.

What Myin and O'Regan (2002) add to Clark's skill theory is another crucial aspect of the way organisms engage with the world. This has to do with intentionality. In order to exercise a particular skill, an organism must pay attention to the world in a particular way. This way of engagement is guided by the knowledge that is embodied in the exercise of a particular skill, which has been acquired through practice and experience. This knowledge is implicit. Humans, for example, cannot describe the knowledge they have that allows them to walk or to run, or the difference between them, though we do have distinctive feelings associated with these modalities of movement. The process of becoming aware of one's motion or perception depends on two factors: the intention to deploy a skill of a particular kind, and its implementation. It is the intention that allows the organisms to engage with the world in a specific manner, so that it pays attention to a subset of the countless stimuli available to it at any moment. Without intention, the organisms would be overwhelmed by the diverse possibilities presented to it. Diversity represents the possibility for appropriate, creative action, but attention needs to be focussed on particular ways of making sense of the world. Without intention, we can see, hear, and do nothing.

Myin and O'Regan argue that their extension of Clark's skill theory to a 'sensorimotor contingency theory' allows us to understand some fundamental aspects of experience that have been identified as 'the hard problem' associated with the experience of qualities such as redness or softness or beauty (Chalmers 1996). In particular, there are two aspects of this experience that seem difficult to account for, one of which is 'ineffability' and the other 'subjectivity.' Ineffability is the property that you can never adequately describe what you are experiencing, what you are paying attention to or are aware of. However, since it is not possible to describe all the knowledge that underlies the exercise of a particular skill, it is the implicit knowledge that is the source of ineffability.

Subjectivity describes the fact that experience is for a subject. Since becoming perceptually aware of something means interacting with it, devoting one's skills to exploring it, the active perceiver is necessarily the centre of the experience. As a result, we have a foundation here for a natural science of qualities that recognizes organisms as intentional agents with direct experience and knowledge of the qualities of what they are paying attention to in the world. This provides us with an approach to the study of feelings and experience through a detailed examination of the ways in which organisms engage with the world. The pig study is fully compatible with these insights: people have direct access to the quality of experience reflected in the behaviour of another organism by virtue of shared modes of perception between humans and pigs. How far this can be extended to other species of organism, especially rather remote species such as ants or plants and our capacity to evaluate their quality of life, can now be more systematically explored.

The emergence of feelings

We can now return to the question how feelings are possible in living organisms. So far we have explored the conditions under which experiences with particular qualities arise, which involve active, intentional agents engaged in probing their world in specific ways that require the exercise of skills. Can we say more about the processes that accompany the experience of feelings? That is, can we identify some form of activity within organisms that strictly correlates with subjective experience? The point of this exercise is to get some idea of what activities are necessary and sufficient for the experience of feeling so that we have some indication of what kind of organized process is required in organisms (or artefacts like computers or robots) for subjective experience to arise. Intentions and skills are certainly necessary as generators of specific modalities or qualities of experience. This may be as far as we can go in defining the generative basis of qualities in terms of the activity of a complex self-organized process: the organism engages in an organized process of exploration of its environment. However, is there a further set of activities within the organism that can be identified as necessary accompaniments to this, giving us insight into the processes that generate experience?

At this point we make contact again with scientific reductionism, giving us a direction in which we can move with this inquiry. However, this is not a reduction of an organism's intention and skilled engagement with its surroundings to a different level of process, but an attempt to get some sense of the universality of experience throughout the living realm. How complex must an organism be to have feelings? In general, it is assumed that it has to be pretty complex, especially with respect to its neural organization. Now I shall describe a proposition that suggests that this may not be the case. Relatively simple organisms may have feelings, in this view.

The starting point of this view is that subjective experiences and their corresponding objective properties are two fundamentally different manifestations of the same underlying reality. One is not more basic and real than the other, as Western scientists have tended to argue. Then subjectivity may be an aspect of any level of organization of process in nature, but different expressions of subjectivity in nature will arise in different types of organized process. This would then be just like different qualities of 'hardness' or 'fluidity' being expressed by different forms of organization of matter and energy. I shall explore now one approach that connects directly with the principles of complexity and emergence that have informed this essay from the beginning.

Some initial focus on this question comes from the recognition that there does seem to be a direct correlation between the activities of nerve cells and the emergence of what we call consciousness, including feelings. What aspect of neural activity is involved in the expression of the type of agency that is connected with experience? An answer to this at the moment can only be a suggestion for further exploration. However, a plausible proposal has been made by Romijn (2002) that takes the investigation in very interesting directions that are consistent with a significant tradition of thinking in this field.

Neural activity is accompanied by electrical action potentials that propagate along neurons and changes in the resting potential of cell bodies. These produce transient changes in electric and magnetic fields that are distributed throughout the neural networks. The quantum mechanical description of these fields of force, which involve attraction or repulsion of the type observed between electrically charged objects of different or of like charge, respectively, or magnetic poles ('north' and 'south') of opposite or like sign, is in terms of virtual photons. These travel between the electric or magnetic poles along trajectories that are described as lines of force, producing repulsive or attractive forces depending on whether the sources of the virtual photons have the same or different signs. These ephemeral photons have been described as joining the sources together in a manner similar to the way in which two tennis players are joined together by the continuous exchange of the tennis ball.

The virtual photons can undergo dynamic processes of organization into coherent states of varying degrees of order, just as occur in all elementary particles of physics. Coherent states of real photons are familiar from lasers, which manifest high degrees of order so that the photons don't scatter and dissipate energy as in ordinary light, allowing us to bounce a laser beam off the surface of the moon and record its passage back to earth. Similarly, virtual photons can become coherent in their activities. Romijn (2002) suggests that it is the degree of coherence of virtual photons in neural systems that is correlated with different levels of subjective awareness. Subjectivity and objectivity are then related to different aspects of order and organization in the same underlying reality. All matter constantly emits and absorbs virtual photons. According to the above suggestion, subjectivity can emerge wherever these virtual photons become coherently organized in particular ways. It may be that the requisite order emerges only in complex nervous systems. However, it is evident that this is not necessary and different degrees of subjectivity could emerge in systems with different degrees of order. Certainly for the type of attentional engagement with the natural world that occurs in humans and pigs and birds a high degree of sensorimotor coordination is required to generate different modalities of experience, assuming that we can extend the human situation in this way to other animals. However, it now becomes possible to see how there could be an emergence of distinct forms of subjective experience at many different levels of order in the biological realm, and possibly throughout nature when particular conditions of organization are satisfied. The particular order that is connected with subjective experience involves active agents with intentionality that engage with the world in particular ways, generating coherent fields of virtual photons in their neural systems that result from specific forms of excitation of the neural networks. Different types of agent engage in the world in specific ways, making them distinct in their modes of being and the type of experience available to them; but there is also a shared or universal aspect to their experience which may be embodied in the coherent states of their virtual photons. This is not an attempt to explain away subjective experience by reducing it to coherent virtual photons. In fact there is no causal explanation being offered, but simply a search for a consistent description of the type offered by complexity theory that could guide further research into an aspect of our world that has resulted in serious divisions and damage to the health of our communities and how we relate to nature.

The suggested extension of subjectivity to basic aspects of nature has been described as pan-psychism or pan-sentience, following the line of reasoning of philosophers such as Whitehead (1929) and Hartshorne (1968) (see Griffin 1998). It transforms the nature of the world in which we live, since what we call 'matter' is no longer 'dead' but has the potential for experience at any level of appropriate organization. However, it remains a territory of extreme disagreement and dispute, and is generally rejected by the scientific and philosophical establishment, which continues to be split by the paralysing dilemma of living with and by feelings and qualities but being uncertain of their reality. I explore this further in Chapter 5.

To live is to know; to be human is to love

All organisms have skills that allow them to engage with their environments and to 'make a living' in species-specific ways. In *The Tree of Knowledge,* Maturana and Varela present a perspective on evolution which argues that every species of organism has specific knowledge of its environment, and that it constitutes the environment that it knows through the intentional and appropriate exercise of these skills. Their proposition is that to live is to know, so that living organisms are characterized by their ways of knowing and being in the world. Since much knowledge is implicit and embodied, this perspective takes us well away from the Cartesian separation of a sensitive, thoughtful mind from its mechanical, automatic body, to a unity of mind and body that joins thought and action, quantity and quality of experience, nature and culture. This is compatible with Darwinism, but it goes much further in providing a basis for understanding organisms as sentient beings that create and know their worlds by exercise of intention and skill in which feelings are genuinely emergent aspects of nature, as described above. Biology has worked hard at reducing the living condition to mechanical processes at the molecular and genetic level which, together with natural selection, are taken to provide a basis for explaining all the phenomena of evolution. This has been immensely fruitful and valuable as far as it goes, but we can now see the serious limitations of this account.

At the end of their book, Maturana and Varela say of human culture:

> We have only the world that we can bring forth with others, and only love helps bring it forth.' So for them, to live as a human is not simply to know and to feel but to love. 'This is the biological foundation of social phenomena: without love, without acceptance of others living beside us, there is no social process and, therefore, no humanness.' (Maturana and Varela, 1987)

This quality of living in relationship is often regarded as the ultimate expression of human potential. The process of human culture is then guided by this quality of living and relating as the full expression of responsible freedom. The present that we know and experience is continuously being formed by our values, which we use to construct our future, to bring into existence the world in which we live as active, feeling agents. Love holds open all the diverse possibilities for participation and creative action by the members of the community, whatever it may be. No-one is isolated or invalidated. Diversity of perspective is acknowledged and valued and conflict provides energy for transformation and resolution. Nevertheless, there are always necessary bounds to action that are negotiated by consensus, creating the constraints that are part of the spur to appropriate creativity.

4. Evolution with Meaning

Separating nature from culture

The natural sciences such as physics, chemistry, and biology are generally regarded as providing a window on to nature, while the arts, humanities and psychology describe the realm called culture. As mentioned in the Introduction, these two areas of learning have been sharply separated in modern thought because of the belief that during the course of evolution properties such as consciousness, language and ethics have emerged in humans, but not, with possibly minor exceptions, in other species. Furthermore, 'meaning' is a term that is used in the context of human language, communication and writing, but is considered to make no sense in the context of developmental and evolutionary processes. After all, these are considered to be based on blind mechanical activities in organisms, which are meaningless. As stated so clearly by Richard Dawkins: 'Each one of us is a machine, like an airliner only more complicated' (1986).

This is the view that prevailed throughout most of the twentieth century. However recent biological studies on molecular and genetic organization within cells are leading to a new perspective on how organisms make themselves which involves a radical rethinking of the distinction between nature and culture. It appears that when developing organisms read their genomes and make sense of them by constructing themselves as coherent, functional wholes appropriate to their environmental contexts, whether they be worms or plants or humans, they are engaged in making meaning through the use of a language of relationships that has deep affinities with spoken language.

Information and the genome

The state of knowledge in biology now, early in the twenty-first century, presents us with the challenge of emergent properties in organisms and their meaning with renewed intensity. Darwin's recognition of the importance of heredity in the evolution of species gave rise to an intense focus on the role of genes in organisms during the twentieth century. This led to the elucidation by Watson and Crick in 1953 of the structure of the DNA double helix and its remarkable properties. The realization that this pair of entwined polymers have the properties required to copy themselves accurately and also to send messages to the cell resulted in a transformation in our understanding of living processes at the most basic molecular level. The result was the molecular biology revolution of the 1960s, 70s and 80s, which was based on the recognition that heredity is about information copying and message transmission. This description of living process transcended physical and chemical principles in the sense that biology entered a new world based on the metaphor that DNA is like a computer programme, organizing all the molecular activities of cells and organisms. Not only does DNA copy itself when cells divide, making faithful replicas so that all its store of information is inherited by the new cell; the information in the DNA is such that it also defines how the macromolecules produced by transcription and translation of genetic information interact with the DNA and with each other in organized sequences of activity in the cell. This revelation fuelled two decades of intensive research (the 80s and 90s) into the secrets of life that became known as the genome project. The underlying principle is that each species of organism has a specific genome defined by its particular collection of genes with the information required to specify everything about that species, making it distinct from all others.

At this point we need to look carefully at the above statement. It claims that the information in the DNA specifies all the characteristics of a species; that is, this information is sufficient to tell us how an organism of a particular species, such as a willow tree or a frog or a human being, is made during its development from an egg. Here is how one proponent of the human genome project described the expected outcome of sequencing and identifying all the genes in human chromosomes:

[The] collection of chromosomes in the fertilized egg constitutes the complete set of instructions for development:
the timing and details of the formation of the heart, the central nervous system, the immune system, and every other
organ and tissue required for life. (Delisi 1988)

What was the evidence for such a strong statement? It was based on
observations of the following type. There are mutations in single genes
that produce remarkable transformations of body form in the little organism called the fruit fly, a favorite for genetic research because of its short
development time (two days) from egg to adult fly. For instance, when a
gene called eyeless fails to act, the result is a fly without eyes (genes are
often described by the results they produce when they mutate and fail to
act). A mutation (in a gene called ultrabithorax) produces a fly with four
wings instead of the usual two, because little structures called halteres of
the normal fly develop into fully-formed wings. Also, it was found that if
a particular gene (called Pax6, a member of the eyeless family of genes)
is switched on in tissues where it is normally inactive, such as the tip of
a developing wing or leg, an eye will form there.

These striking observations on the power of changes in single genes
to alter the morphology of an organism is what led many biologists to
take the view expressed by Delisi: that by reading the information in the
genes of an organism (its genome) we can understand everything about
that organism, including how it develops from the egg. However, even
before the genome project got underway with the promise of revealing all
the secrets of the functioning of organisms, and in particular embryonic
development, there were sceptical voices saying that this position is more
rhetoric than reality, based as it was on an incorrect interpretation of the
genetic evidence. Take the case of the eyeless gene. The genetic observation is that if this gene fails to act, no eyes form. Compare this with the
simple observation that if you remove the sparkplugs from an engine, it
won't work. Remove other parts, such as the battery or the distributor or
the carburettor and again the engine won't work. Would you conclude that
a knowledge of all the different parts is the key to understanding the internal combustion engine? They certainly are important. But understanding
what they do, and understanding what all the other parts contribute to the
functioning of an engine, won't give you the answers to how the engine
functions as a whole. You need the plan of construction, how the parts

are organized in relation to each other, and a lot more knowledge about combustion and electricity and the physics of gases to understand how an engine works. What underlies the view of Delisi and many others who worked on the genome project is the assumption that organisms are sufficiently like the machines we construct so that a knowledge of the parts would allow us to work out how they operate. This is the assumption that has turned out to be radically wrong. And there were many who knew this long before the genome project began. Because even the genetic evidence of several decades ago told us that genes interact in very complex ways. Other genes can change to compensate for the failure of any particular gene so that, for instance, flies can make eyes even when the eyeless gene is inactive. This has been known since the 1940s and is a commonplace of genetics. (cf. De Beer 1971; Webster and Goodwin 1996). It is extremely rare for any single gene to be solely responsible for a particular function in an organism. Only when a gene codes for a single molecule that is itself the cause of a function, as in the case of an enzyme that catalyses the transformation of one molecule into another, can it be said that one gene causes one function. Even this situation is now known to be subject to modulation, because the protein can be altered by attachment of different chemical groups at different sites in the protein after it has been formed, modulating its function in specific ways.

Despite this evidence of complexity and subtlety in the working of genes, the idea that single genes are the primary causes of complex processes, such as the formation of eyes in the fruit fly, continued to inspire many participants in the genome project with the belief that sequencing the genomes of different species would reveal all their secrets. The Pax6 gene provides a good example of how this thinking persisted into the 1990s. As mentioned above, when this gene is activated in developing tissues that would not normally make an eye, such as those developing into a wing or a leg, an eye can appear. This gene was described as a 'master gene' for eye development by the developmental geneticists, Walter Gehring and his colleagues (Halder *et al* 1995; Gehring and Ikeo 1999). It was proposed that this gene initiates eye formation not only in the fruit fly but in all bilaterally symmetric organisms with eyes, which are descended from a common ancestral form millions of years ago. The discovery of genes that have been conserved and serve similar functions in different species within a common evolutionary lineage was one of the great insights of developmental genetics in the 80s and 90s. It made profound sense that when new genes arose that provided organisms with useful functions, these genes would be selected and conserved in their activities so that they didn't have to be rediscovered by species that emerged later. Gehring's master gene hypothesis was simply a very strong and specific formulation of this idea, in which the gene involved had an identified function, to initiate eye formation. If true, it considerably strengthened the view that finding out the functions of the genes of a species by sequencing them and identifying the information they contain would lead to the understanding of development that Delisi promised.

As expected by those who recognized that all genes act within a context not only of other gene activities but of a cellular environment that influences the effects of genes and their products, the master gene hypothesis has been found wanting. Pax6 does not act on its own but cooperates with several other genes in tissues which are competent to form eyes in a developing embryo; and it is unable to initiate eye development in other tissues that lack this competence (cf. Wilkins 2002). The lesson that has emerged with blinding clarity from the whole genome project is the error of regarding an organism as a kind of supermolecular machine whose parts are written in the genetic code. Since we learned to decipher this half a century ago, it was plausible to proceed on the assumption that all we needed to do was read the messages in the genes to understand how this machine is made. Science makes progress by engaging in clearly defined, tractable projects like this. These projects are described as convincing stories about issues that are topical both within the scientific community and in society generally, since much of the research funding for these projects comes from the public purse. The story about DNA as the secret of life, which has been told to society since the 1950s, makes extremely good reading and is one of the great scientific insights of the twentieth century. When combined with the story that we can now read this secret by genome sequencing, and hence understand life in predictable detail, we had a thrilling sequel to the drama of the double helix as the heart of living matter. Stories always have elements of truth combined with imaginative and plausible elaboration. They are also filled with values and moral lessons, and scientific stories are no exception. The justification for the genome project was not simply the insight we gain into how organisms are made and how they function, but also the benefits to humanity from control of virus and microbial parasites, alleviating genetic diseases in humans, and manipulation of plant genomes to improve food crops, to name but a few. The fact that these promises now look distinctly more difficult to deliver, and the issues raised by civil society about the value of some of them, such as genetically modified food, does not alter the fact that stories continue to be the vehicle for social dialogue. One story leads to another.

Changing the story of the gene

Evelyn Fox Keller, reflecting on the outcome of the human genome project just before its completion in 2001, commented:

> What is most impressive to me is not so much the ways in which the genome project has fulfilled our expectations but the ways in which it has transformed them. ... Contrary to all expectations, instead of lending support to the familiar notions of genetic determinism that have acquired so powerful a grip on the popular imagination, these successes pose critical challenges to such notions. Today, the prominence of genes in both the general media and the scientific press suggests that in this new science of genomics, twentieth century genetics has achieved its apotheosis. Yet, the very successes that have so stirred our imagination have also radically undermined their core driving concept, the concept of the gene. As the human genome project nears the realization of its goals, biologists have begun to recognize that those goals represent not an end but the beginning of a new era in biology. (E.F. Keller 2000)

Why is Keller suggesting that the concept of the gene has not survived the genome project? The first shock was the realization that humans, generally regarded as the most complex of organisms, do not have many more genes than, say, a tiny plant called mouse cress. The plant probably has about 27,000 genes while we have somewhere around 30,000. It is not the number of genes, and hence the quantity of genetic information that is important in determining the complexity of an organism, but how the genes are organized, their relationships. This was clear when it emerged that the difference in gene number between humans and chimpanzees was no more than 2%, but it was still expected that humans (and chimps) would have around 100,000 genes, coming out way ahead of plants. The result of the genome project on expected gene numbers in different species was a shock to the gene industry.

Another of the recent revelations that has sounded the death knell of the traditional gene concept is the result that a sequence of DNA that was once regarded as a gene that coded for a single protein can in fact be read by a cell in hundreds of different ways, each one resulting in a different protein. For example, in the hair cells of the inner ear of the chick there is a gene that can be translated into 576 different proteins, each one altering the tuning of cells to sound frequencies (Black 1998). This is possible because the single sequence of bases in the DNA that contains the information for making proteins can be read in many different ways by what is called alternative splicing. Sections of the genetic sequence from the 'gene' are cut and joined in a variety of ways to produce different messages that make different proteins. We do not know what determines which of these many variant proteins is produced in hair cells in different positions of the inner ear, but it is recognized that the translation processes involved are exquisitely sensitive to context. In the fruit fly, it has been estimated that the number of different messages that could arise from a single 'gene' sequence is 38,016! Faced with such overwhelming complexity, the question arises what lessons we have learned and how we are now to proceed. What story can we tell about organisms and their genes that will inform useful research?

Keller suggests that a major lesson from the genome project could be a qualitative shift in attitude on the part of the scientific community:

> It is a rare and wonderful moment when success teaches us humility, and this, I argue, is precisely the moment at which we find ourselves at the end of the twentieth century. Indeed, of all the benefits that genomics has bequeathed to us, this humility may ultimately prove to have been its greatest contribution. For almost fifty years, we lulled ourselves into believing that, in discovering the molecular basis of genetic information, we had found the 'secret of life'; we were confident that if we could only decode the message in DNA's sequence of nucleotides, we would understand the 'program' that makes an organism what it is. And we marveled at how simple the answer seemed to be. But now, in the call for a functional genomics, we can read at least a tacit acknowledgement of how large the gap between genetic 'information' and biological meaning really is. (E.F. Keller 2000)

Looking for meaning in biology

'Humility' and 'meaning' are not words that are generally found in
scientific research programmes. 'Functional genomics' is much more
what one expects, a technical-sounding expression that suggests a
way forward from genes to the manner in which they function in
the organism. The challenge for biology at the moment is to adopt
a new metaphor that can carry forward our quest to understand life
with a research programme that has continuity with the past but also
moves decisively into a new mode of action, perhaps also involving
both humility and the search for meaning. The results of the genome
project can be seen as a major wake-up call in biology of the type
that occurs periodically in the history of science, such as the ones
described in Chapter 1: relativity theory, quantum mechanics, chaos,
complexity. It is another part of the story of the Prince waking up to
the realization that he can accept the world as an ambiguous, magical
place where things are not always what they seem. Hence the need for
humility. At the same time, we live our lives in terms of meaning. So
what if we actually take this to be an aspect of organisms themselves:
they live lives of meaning. How do they do it? As Stuart Kauffman
(2000) makes clear, organisms don't compute their lives in the manner
that the information/machine metaphor suggests, they *live* them. What
could this mean? Here I begin a new story that explores this question,
and I start with a fairy tale.

The beekeeper's son

Once upon a time there was a woman who kept bees. She lived
on the outskirts of a little village and the villagers all said that the
honey from her hives was the sweetest and most fragrant they
had ever tasted. The honey was also reputed to have remarkable
healing properties, both for cuts and bruises and, taken with
various herbs as a drink, for all kinds of illness. It was even said
that the honey kept old people strong and vigorous long after
the age when most people begin to lose their energy.

The beekeeper loved her bees and cared for them nearly as
much as she did for her husband and her one son. Her husband

wasn't very interested in the bees but her son was, and he learned how to look after them from his mother: when to give them new frames for their honeycombs as the colony grew; how to take out the frames that were full of honey without disturbing the bees; when the bees needed extra food in the winter months; when the bees showed signs of swarming and whether to start a new hive or to return the swarm to their original home. But while the beekeeper just seemed to know what the bees wanted and how to tend to all their needs, her son became more and more interested in knowing how the bees managed to create such a perfect harmony within the hive and in relation to the seasons and the countryside where they foraged. He wanted to know how the bees found the scattered sites where different types of flowers bloomed at different times of year, how they found their way back to the hive, how they told their fellow workers where the sources of nectar were and what quality it was, why there was only one queen in the hive when many newly-hatched larvae had the potential to become queens, how the workers decided which one of the larvae to feed specially so that it became the new queen, and many other questions. So he studied them closely. He was sure that by watching and paying close attention to their behaviour he could discover all their secrets. Above all, he wanted to know the meaning of the life of bees so that he might get some insight into the meaning of his own life and how he should live it.

One very hot day after he had given the bees extra sugar water because of the heat wave, he sat beside a hive to see how they coped with the high temperature. Would they forage as usual, or stay at home? As he watched and listened, he became aware of a sound he hadn't noticed before. At first it was just a distant hum, but then he realized that the sound was a kind of purring, many different notes all blending to create a sound so rich and full that it was like the voice of love itself. He realized that the bees in the hive were fanning themselves, thousands of wings beating in harmony to create a delicious draft of air that kept them all cool. As he listened, the beekeeper's son fell into a reverie in which he himself became a bee within the hive, taking part in the cooperative creation of the colony. Totally immersed in the delicious sound, he began to hear a single unified voice that sang over and over again: this is the hive of bee-ing, this is the hive of bee-ing, this is the hive of bee-ing.

The beekeeper's son suddenly awoke from his reverie with the realization that the bees had given him the answer he sought. He had been asking the wrong question about the meaning of the life of bees, and the meaning of his own life in the scheme of things. He turned the question around to ask not 'What is the meaning of life?' but 'What is a life of meaning?' And the answer from the bees was: meaning is not what you do and how you do it, but in the very act of being in harmony with your nature and those of all other beings. Henceforth the beekeeper's son stopped looking for answers to all his questions and, whenever he wasn't sure what to do, he hummed the bee song to himself until he felt the right action in his heart and then followed it.

Meaning and language

Suppose we take the view that an organism reads its genetic script and makes meaning out of it by expressing the script in its form, by which I mean its morphology and behaviour? Then we might say that an organism, starting as a fertilized egg, makes meaning of its genetic inheritance by using it to make a coherent being, the functioning form of the organism. This not quite that same use of the terms 'read' and 'meaning' as when we talk about someone making meaning from a written text through the act of reading. A person makes meaning by telling a coherent story or by making sense of the text. The vehicle is language. If the metaphor of organisms making sense of their genomes by reading and expressing them in living forms is going to work and be useful, then we need to examine language and meaning to see if this new reading can give us new insights into the way organisms make themselves.

How do you find out the meaning of a word in a language? You look it up in a dictionary. And what do you find? You are given other words and phrases in that language that express the meaning of the word. This can be a frustrating process because you may have to look up some of these other words until you find ones with which you are familiar so you can make sense of the dictionary entry. For instance, I look up 'meaning' and find: 'the sense or significance of a word, sentence, symbol, etc.' Clearly I have to be familiar with a language, to be a practised language user, since the meaning of a word is not something that exists in itself but arises out of its relations or associations with other words. The recognition of this fundamentally relational aspect of language and its systematic study was the foundation of a movement in European thought called Structuralism. This started in linguistics and moved into a great diversity of related aspects of human cultural activity including science, in particular biology, as we shall see. The originator of this twentieth century movement was a Swiss linguist called Ferdinand de Saussure. He described language as a set of elements mutually related to each other in such a way that the state ('meaning') of each element determines and/or is determined by the state of other elements, and every element is connected to every other by a chain of such determining relations. In language the elements are words, which relate concepts to sound-images, the combination of which de Saussure called a sign. The psychological, experiential basis of language was thus acknowledged while its formal structure was examined in structuralism. Words as signs signify something in experience, that which is signified. The sign or word is arbitrary in the sense that there is no necessary relation between the concept and the word spoken, although once a word belongs within a language it cannot be arbitrarily changed. The crucial property of signs is their ability to distinguish between concepts, so that it is their differences that matter. Any system of signs could be used so long as they make distinctions between different concepts. Our organs of perception and of thought also fragment or separate the world of appearance in more or less arbitrary ways, in that different organs could have separated it in different ways, as they do in different species. The striking property of language, which was discovered by structuralism, is that the association of two contingent systems of difference is capable of producing a system of significance within which we can talk of meaning (see, for instance, Caws 1988).

There is an important distinction made in structuralist linguistics between the formal structure of language as a system of rules and relationships, defining a syntax and grammar, and everyday speech, which is spoken by persons and is individually distinctive. De Saussure used the French word *langue* for the general system of norms whose internal structure and workings he was concerned to study. For spoken language, the utterances of an individual in a particular context, he used the word *parole* (literally 'spoken word' or 'discourse,' understood as an activity rather than an object).

What, then, of meaning? Caws (1988) describes a structuralist theory of meaning as the product of two more fundamental processes which he calls signifying and mattering. 'Signifying is the defining property of signs and the fundamental mark of the intelligible; mattering is the defining property of value and the fundamental mark of the purposive.' So when we speak of meaning we are talking about both making distinctions in experience and communicating these, as in language; and we are referring to the values or the feelings associated with an intended action, why it matters to us. Caws expands on this in a very interesting way:

> The roots of mattering lie in the structure of biological needs, and it is in the intelligent satisfaction of those needs that meaning first comes into play. In the case of the matching of a structure of praxis with a structure of desire the proportion of the purposive to the intelligible is very high; as the immediacy of need lessens, the balance of the two components changes, until at an advanced stage of culture or education a very complex structure of intelligibility may be evoked by almost casual purposes. One of the ways in which the meaningfulness of significant activity is maintained is through an intention to pursue the intelligible for its own sake, and in high civilizations this becomes, in literature, music, art, and other forms of creative activity, the dominant exercise of meaning. (Caws 1988)

Here we have a way of rethinking the whole of biological and cultural evolution from a perspective that involves simultaneously the intelligible or rational and the affective or experiential, putting thought and feeling together at the most fundamental level of understanding what it means to live a life of meaning. The first step is to rethink the organism as a being that simultaneously possesses a language whereby it makes sense or meaning of its genetic text and engages in purposive activity to satisfy its needs. The second half of this view is already familiar from Darwin's functional perspective: organisms are adapted to particular habitats within which they 'know' how to make a living. Knowledge now includes inherited patterns of appropriate behaviour, or instincts. It can also be used in relation to the structures within the organism whereby such patterns arise, as in the structure of the 'language organ' in humans, which can be seen as an embodied form of knowledge (Chomsky 1968, *Language and Mind*). Such a perspective is developed in detail by Maturana and Varela in their book *The Tree of Knowledge*. In describing organisms as autopoietic (self-making) agents, they argue, as we saw earlier, that to live is to know: organisms function by virtue of embodied, inherited knowledge. This description applies to the simplest of creatures, such as bacteria, which 'know' the difference between sugar and sulphuric acid, swimming towards the first to get a meal and avoiding the second to prevent meltdown. Kauffman (2000), who uses the term 'autonomous agent' for living beings in a manner similar to the 'autopoietic agents' of Maturana and Varela, describes the primordial microbial community of three billion years ago as follows. 'I want to say that the autonomous agents comprising that community had, individually and collectively, the embodied know-how to get on with making a living in the natural games that constituted their world.' This view is greatly expanded and given a firm empirical grounding by Ben Jacob *et al* in a discussion of cognition and social intelligence in bacterial colonies. Here is how they describe their observations on the behaviour of colonies under different circumstances of nutrition and interaction:

> The term 'cognition' usually refers to human mental functions associated with capacities such as the use of semantic and pragmatic levels of language, perceiving self vs. non-self, association with group identity and perceiving individual and group goals. It is now realized that bacteria facilitate surprising collective functions. They can develop collective memory, use and generate common knowledge, develop group chemical identity, distinguish the chemical identity of other colonies in their environment or even higher organisms, learn from experience to improve their collective state and more.
>
> These are the bacteria faculties we refer to when using the term fundamental elements of cognition. We emphasize that these features should not be confused with the unique,

human level of symbolic cognition. We do not imply that
bacteria possess human capabilities but that fundamental
elements of cognition can also be found in bacteria. From a
practical perspective, this realization can shed light on the
evolution of cognition and on the most basic requirement
for its facilitation in all organisms.

By meaning-based natural intelligence we refer to that
common trait shared by all organisms, from bacteria to
humans, which share some fundamental cognitive functions such as sensing, information processing and contextual interpretation of information (the ability to assign
contextual meaning to externally gathered information). By
contextual, we refer to how external latent information is
placed within a framework in which its relevance (meaning), is derived according to the organism's external and
internal conditions and internally stored information. The
ability to assign contextual meaning to externally gathered
information is a fundamental semantic function of natural
intelligence that every organism
must have. (Ben Jacob *et al* 2006)

Seeing organisms as agents that intelligently satisfy their needs by
means of embodied knowledge is a view that has arrived, science having
now caught up with what most people recognize intuitively. The implication is that an organism is a member of a community (a species) with a
shared language and a culture that makes meaning of its history (reads
the inherited text) by creating forms that are appropriate to their context.
This view places organisms firmly within a perspective of creative culture that we recognize as one of our own aspirations. We see intuitively
that species live meaningful and appropriate lives, and we seek the same
condition. As has been remarked, all species on earth seem to know their
jobs except humans. We as a relatively young species are on a learning journey whose outcome is in our hands since we have the freedom
to make choices, as have other species. So a perspective that seeks to
articulate a view of life with intrinsic meaning may be of value not only
for understanding organisms, but also in seeking our own place in a
world from which we have alienated ourselves as a result of a particular
cultural exploration.

Hermeneutics

The study of texts and their meaning is called hermeneutics. Hermes is the messenger of the Gods, and it is to him that we turn in order to make meaning of the book of life. Hermeneutics was first developed in connection with the interpretation of sacred texts such as the Bible or the *Koran* or the *Bhagavad Gita,* but it is now used generally as the 'science of interpretation,' as my dictionary puts it. It studies the process whereby readers make meaning of texts through their interpretation of them in the context of the reader's own culture and place in history. Applying this to organisms to get a hermeneutic biology, we need only rewrite the description as: a study of the process whereby organisms make meaning of their genetic texts by expressing them in a form (morphology and behaviour) appropriate to their habitat and their history. Modern science is often described as the process whereby scientists read the book of nature to discover her laws (Galileo, Bacon) and hence to grasp the intelligibility of the natural world. It has been assumed that there is only one true reading of nature in this sense, so that science is a single coherent unfolding of the truth. This has given rise to the belief that there is only one scientific story that is consistent with all the facts, which is the grand scientific story or metanarrative continuously sought by the scientific community. This position has come under heavy fire because of the very nature of scientific exploration, which proceeds without any need to have consistency overall. The different branches of science tell their own stories so that biology proceeds without reduction to physics, hydrodynamics proceeds without reduction to quantum mechanics, quantum mechanics works without needing to be integrated with relativity, and so on. It is generally assumed that nature is self-consistent, and science often demonstrates this in achieving unified descriptions of diverse areas of study. Putting quantum mechanics and relativity together into a unified theory is the research project called quantum gravity, one of the most exciting frontiers of science. But each of these subjects works very well within its particular domain of explanation, telling locally consistent stories.

It is now realized that theories in science are primarily local to a discipline, and that a theory of everything is largely empty of meaning because of its extreme abstraction. Hence science provides different readings of nature that need not be consistent, simply coherent within the discipline.

This is the realization of postmodernism: an overall metanarrative is a chimaera that will never be realized because of the very nature of science as a creative, always unfinished activity. Furthermore, if there were such a grand, objective synthesis it would have the property of existing independently of us, who simply work away at the understanding of this integrated whole that is objectively given. This is indeed the position adopted by many scientists, and it continues to animate the search for unified theories (cf. Taylor 2001). However, the result is an absence of the subject and its qualities from the very fabric of the cosmos, presenting us with logical contradictions in trying to account for how it is that subjects with feelings can emerge from matter that has no such properties. These difficulties were discussed in the last chapter, where an alternative approach was explored that required some fundamental changes in the nature of 'reality.' Changes in science don't come one at a time but in flocks, because of the interconnected nature of the whole enterprise of making sense of the world we belong to. The change we are now contemplating is a shift from scientists as observers of a given world to scientists as co-creators of that world with beings that are much more like us cognitively and culturally than we have hitherto recognized. 'The history of science is science itself,' as Goethe observed. We are within the history that is unfolding.

The objective stance of modern science, with the observer standing outside and studying the principles of order in a given cosmos, is the position that results in science being amoral, with matters of fact being seen as independent of values. The world just happens to be as it is, and it tells us nothing about how we should behave, it is claimed. However, this changes when the subject is put back into the world as a participant in its process. It is the subject who chooses what to study and how to make sense of the world, so the structure and content of science is a result of how we proceed with our investigation. This does not mean that scientific theories are simply subjective constructions without objective legitimacy, as some postmodern authors would have it. What science seeks to understand ('nature') is not our creation, so scientific truth still refers to something real and science continues to subscribe to realism. However, it is necessary to recognize that what we call 'objective' in science is consensus within a community of subjects who agree to a particular methodology of investigation. This is what gives science its democratic, communal, consensual basis. But the community of scientists is diverse in the details of its methods of examining nature, and in the theories that arise from these. So a unified enterprise with the objective of understanding the world as a consistent unity necessarily breaks into many different sub-communities working on different aspects of the world, seeing local consistencies and assuming global consistency but never actually arriving at the overall picture. Paradoxically, the modernist attempt to structure our existence in terms of objective absolutes leads to nothing less than our imprisonment in an amoral enterprise, while the postmodern attitude sets us free, not to do as we like, but to behave ethically.

Recognizing the local, communal aspect of scientific activity as subjects engaged in seeking consensus about their work puts an end to the idea that science can ever produce a cosmic metanarrative, with the consequence that ethics returns intrinsically to scientific activity. The reason for this is spelled out by Cilliers as follows:

> It becomes the responsibility of every player in any discursive practice to know the rules of the language game involved. These rules are local, that is, limited in time and space. In following such rules, one has to assume responsibility both for the rules themselves and for the effects of that specific practice. This responsibility cannot be shifted to any universally guiding principles or institutions — whether they be the State, the Church or the Club. (Cilliers 1998)

Science itself now becomes a hermeneutic activity, like the rest of culture. The world has many different 'readings,' as we see in the local theories of different branches of science, each of which attempts to be self-consistent.

We can now begin to appreciate what a hermeneutic biology might mean and how it could guide our thinking about living beings. The text that organisms read is their genetic text, which has many different 'interpretations,' each one sensitive to context. This means that there are many different phenotypes that can be produced from a single genotype. The classical evidence for this is to take a plant, make clones of it from cuttings, and grow these cuttings in different environments (for instance, at different elevations but in the same soil). The result is a variety of phenotypes, of forms, that can be so diverse they look like members of different species. This diversity of possibilities is now recognized at the molecular level, single sequences in the DNA (conventionally, single genes) giving rise to many different proteins through alternative splicings, each protein carrying out a slightly different task that is appropriate to its context, as we saw in the gene for the proteins involved in tuning the resonant frequency of hair cells in the inner ear of the chick. We then face the question: who or what are the readers that produce this diversity of interpretations of a single text? There are many sources of molecular diversity of reading, and it is now the task of genomics and proteomics to try to make sense of the way this is done in cells. As stated by Jeong:

> Proteins are traditionally identified on the basis of their individual actions as catalysts, signalling molecules, or building blocks in cells and microorganisms. But our post-genomic view is expanding the protein's role into an element in a network of protein-protein interactions as well, in which it has a contextual or cellular function within functional molecules. (Jeong *et al* 2001)

This raises the level of dynamic complexity in cells dramatically.

Power laws and networks

Just as the level of complexity in molecular networks is producing some bewilderment in researchers, there appear to be structural principles emerging that provide insight into the organization of these networks. Suppose we ask: what is the probability that a particular protein interacts with k other proteins, where k can take any value between 1 and 20? The answer for the proteins of a yeast or bacterial cell is that the probability follows an inverse power law relationship $P = ak^{-\alpha}$ where a and α have specific values. Similar relationships hold for written texts. If we ask what is the probability of finding the *rth* most frequent word in a text, we find that it is $P = cr^{-\beta}$ (where c and β have particular values, generally different from a and α above). This relationship is known as Zipf's law after the name of the Harvard professor who discovered it in 1949. These power law relationships are found in many natural networks with the self-similar or fractal properties described visually in Chapter 2 such as river basins or the branching pattern of trees and the vessels of the circulatory system. This gives these structures certain significant properties, a primary one being resistance to failure or robustness to disturbance. The reason is that most elements in these nets connect to relatively few other elements, so the whole network is not particularly sensitive to a failure in any one. However, there are a few elements that play a central role, being connected to many others, so the system can fail if these are damaged. The robustness of organisms to disturbance and the robustness of texts to random damage may both arise from the way their networks of relationship are organized. Another example of the power law principle is seen in the dynamics of the heart, considered in Chapter 2. The pattern of interbeat intervals in electrocardiograms follows a principle of fractal structure in which the probability of an interbeat interval of size k is $P = ak^{-\alpha}$. The dynamics of the heartbeat is again robust to disturbances from inside or outside the body. What we seem to be encountering here is the possibility that much of the dynamic and relational structure of living systems may follow similar principles, which are emerging from the study of complex systems (cf. Bak 1996; Solé and Goodwin 2000).

Power law distributions characterize the organization of networks that have emerged spontaneously, expressing a type of order that arises from their interactions. The paradigm case for this is what in physics is called a phase transition: the emergence of order at critical points, as when a gas reaches its condensation temperature and begins to form a liquid; or a magnetic material cools to the point where its little magnetic dipoles start to align with one another, the whole becoming a magnet. These are the cases where power laws were first discovered. As will be described in more detail in Chapter 5, they describe situations in which the same pattern is observed at any scale of observation. As new order emerges, the interactions between component parts are independent of what size you use to define a part, so all scales undergo the transition simultaneously. Such processes are called scale-free. Because parts of different size have similar properties, power laws describe features that are self-similar: parts and wholes resemble each other.

Precisely why self-organizing systems exhibit these diagnostic characteristics is the focus of considerable research. While in physical systems these transition points are rare, it seems that living systems express this type of order as a deep aspect of their dynamic nature as continuously self-organizing processes. This has given rise to the conjecture that life occurs on the 'edge of chaos,' since it is frequently the case that living processes have the characteristics of systems poised between a disordered condition and a highly ordered one, as in a physical system at a critical point. One might expect this to be a rather fragile, delicate state. However, it has been proposed that this dynamic condition is in fact a natural attractor for complex self-organizing systems, called self-organized criticality (Bak 1996). That is, such systems naturally fall into a dynamic condition that is characterized by scale-free, self-similar patterns. Solé and Goodwin (2000) demonstrate that many aspects of biological processes, from viral epidemics to taxonomies, the growth of cities and species extinctions, obey power law patterns. Buchanan (2000) has explored the application of these ideas to the description of physical, biological, and cultural processes. While Barabasi sees a whole new science of networks emerging from these and related studies:

> Most systems displaying a high degree of tolerance against failures share a common feature: Their functionality is guaranteed by a highly interconnected complex network. A cell's robustness is hidden in its intricate regulatory and metabolic network; society's resilience is rooted in the interwoven network of financial and regulatory organizations; an ecosystem's survivability is encoded in a carefully crafted web of species interactions. It seems that nature strives to achieve robustness through interconnectivity. Such universal choice of a network architecture is perhaps more than mere coincidence. (Barabasi 2003)

There is accumulating evidence that living processes express a characteristic dynamic state. This provides a direction for research to follow in mapping the connectivities of intracellular networks to see if they obey power laws, as in the work of Jeong *et al* (2001) on protein networks. This is pursued further in the next chapter, but here we need to ask how the hermeneutic metaphor may help us to rethink biological questions from a new and useful perspective within which the sorts of observations about power laws and networks makes sense.

Hermeneutic biology

Someone who has thought long and hard about this question and has written a book about it is Anton Markos, a professor at Charles University in Prague. Originally trained as a biologist, he now describes himself as a theoretical biologist working on the articulation of what it means for organisms to read texts and make meaning of them. Here is what he says in his book *Readers of the Book of Life; Conceptualizing Evolutionary Developmental Biology*:

> The main objective of hermeneutic biology should be to get rid of the genocentric view that takes the genome as a recipe for building the body. It should pose questions about the *builder,* who takes the genome as a mere dictionary of the language in which the recipe is written. Proteins — 'words' uttered in the language — enter into complicated syntactic and semantic relations, which constitute the cellular *parole.* The cell is thus a materialized *parole.*
>
> From this viewpoint, we can pose methodological questions about what understanding can be gained from the contemporary genomic era, which is witnessing the acquisition of complete sequences of genomes. The genome is understood as a complete dictionary and source book of a cell, an individual, a species. Comparative analyses will undoubtedly lead to a deeper understanding of the richness and evolutionary position of a given language. But now another step is needed: to proceed from the dictionary to the utterances. The field of grammar is being mapped very extensively — for example, regulatory and metabolic networks. A new move in the field is the interconnection of gene dictionaries with protein utterances. The semantics of the new science is no longer a Cinderella, even if it is still hidden under names such as 'mechanisms of regulation of gene expression.' (Markos 2002)

The first result of this orientation is a significant shift of focus from the genome to what Markos calls the builder of the organism. We must be careful not to split the organism up into new parts, as happened with

the genotype/phenotype distinction. The builder is not a separate entity
in a cell; simply a new focus of attention. It is the dynamic process that
reads words from the genetic dictionary, producing proteins that are
appropriate to context in the development and functioning of the organism. The suggestion is that this has a structure, a syntax and a grammar that we need to understand as emergent properties of the complex
interacting molecular webs that are called genetic regulatory networks,
signalling cascades, and metabolic networks. Markos describes this shift
in the following terms:

> Cells have active access to their genetic thesaurus: they
> select from it and interpret it in an unceasing confrontation
> with their own time and space (i.e., coordinates within the
> tissue or organism) and with inputs from their environment.
> This work with the 'source code' is dependent on the quality
> of the text itself and the 'tuning' of the cell (the above-men-
> tioned coordinates, physiology, morphology, and history)
> and is indeed to be viewed as a hermeneutic task. A medium
> for this interpretative work, the 'search for meaning,' is to
> hand in the form of non-local morphogenetic fields, con-
> centration gradients, complicated dynamic networks of
> macromolecules (extracellular matrix, cytoskeleton, nucle-
> oskeleton) and morphological structure of longer duration.
> DNA is thus far from being the algorithm prescribing how
> the body will look and how it will behave. It is a genuine
> text to be read by an *informed* (or better, initiated) reader.
> The cell, the body, is an in-formation of experience.

How is this large-scale, coherent regulation of gene activity achieved
in cells? There are several clues to the type of order that may be involved
in gene regulation, one of which was provided by experimental studies
by Guelzim *et al* (2002) on the genetic regulatory network that functions in yeast cells. They show that a design principle of these networks
that sustain cellular functions is a now familiar power law relationship
between the regulating proteins and the regulated genes. We need now to
understand the meaning of such a relationship: what is it telling us about
the evolutionary processes that generate these relationships, if anything?
I shall approach this from a perspective developed by Cancho and Solé
(2003) that gives an important insight into the possible reason for and

significance of the power law relationship. They show in a model of the evolution of language that the tension between speakers and hearers, characterized by the effort to understand the meaning of words in a language, is resolved by a principle of least mutual effort, defined in terms of entropy and negative entropy functions. At the point where this function is minimized the probability distribution of words in a language satisfies Zipf's law, described earlier. The power law arises from a sharp phase transition in the use and assignment of words to objects in the language. The authors conclude that this generic property of natural languages can be understood to arise from conflict resolution expressed in a form similar to the familiar principles of energy or action minimization in physics, which describes states of order such as soap bubbles, vortices in liquids, the patterns of crystals, and so on. The implication is that wherever there are communication networks that are seeking states of minimum mutual effort, power laws will be found as descriptors of properties of the network.

If we now apply this argument to the transcriptional regulatory network of yeast, it follows that we may expect to find a power law describing the relationship between the signals and targets of the communication system, of the type observed by Guelzim. However, there is a further important consequence of this that begins to illuminate the current confusion about the behaviour and properties of the elements of this network. In language, the power law distribution of words is a necessary condition for a deeper and highly significant property of natural languages: the ambiguity of meaning in utterances. Machine languages attempt to resolve this ambiguity by strict one-to-one assignments of 'words' to 'objects.' The result is a form of communication that is mechanical and unambiguous, but utterly lacking in creativity. The creativity of natural languages resides in the multiple meanings that can be assigned to the same sentence because of ambiguities, arising from a non-mechanical or fluid relationship between words and objects in the world. Similarly if the living state is to be robust and creative in its expression of meaning, which is the form of the cell or organism with its adaptable behaviour and functional morphology, then the 'words' of its languages must also be ambiguous so that different combinations can mean the same thing, overall. Hence the regulatory networks must necessarily be ambiguous, not mechanically causal with fixed patterns of interaction. We may expect to find that each cell of a particular type will have a somewhat different spectrum of molecules that correlate with the cell type, though each of the different possible combinations of gene products will obey a power law. This is the type of prediction that my colleague, Philip Franses, and I are suggesting for the frequencies of regulatory signals in cells (Goodwin 2006).

Where, then does the order of the cell come from? It is not mechanically reducible to the molecules that make it up, but arises from a coherent relationship among the molecules at a higher level that in science is usually called a field. Just as in language an intricate pattern of words is used to express a simple, clear meaning, so in a cell an intricate pattern of molecules gives rise to a coherent condition of the cell as it undergoes differentiation to a particular state. This process is irreducible, the set of possible choices in the network resolving into simple complexity that is sensitive to context. It is like the situation in quantum mechanics in which multiple possibilities coexist before being resolved into simplicity appropriate to context. The principle of least effort describes a condition in which the parts of the system and the overall coherence or the order of the whole coexist as necessary partners. The condition is one in which there is simultaneously maximum freedom to the parts (ambiguity of meaning in the relationships between molecules) and maximum order or coherence to the whole. This is the organized process of continuous, robust creativity that characterizes all life, and it is as natural as the production of a soap bubble.

The development of an organism with specific form and behaviour is the act of meaningful creation that the zygote, the fertilized egg, engages in. There are deep similarities between this process and the reading of a text by a reader. In engaging with a well written and well structured text, the reader does not have to wait until the last word is read to grasp the meaning. The whole meaning is present in the first sentence, but the meaning becomes progressively clearer as the text is read. In the same way, the whole organism is present in the fertilized egg, but the specific nature of the organism, its embodied meaning as form and behaviour, becomes clearer as the developmental process proceeds. Like all creative acts, these processes are sensitive both to context and to history, cultural in the case of a reader, environmental and evolutionary in the case of the organism. Hermeneutic biology begins to bring about a convergence of these processes so that living nature and culture are seen to be expressions of similar activities whereby agents create relevant meaning.

The hermeneutic reframing of the nature of the organism involves a reunion of development and evolution within a single process of creative emergence of life cycles. There are different types of dynamic process that can be distinguished in evolution, but they can no longer be separated and put into distinct boxes, such as genetics, and development and evolution. This is already recognized within biology by the emergence of the subject generally referred to as evo-devo, and the appearance of a journal dedicated to developmental evolutionary biology. These new patterns of thinking and their implications will be examined in some detail in Chapter 5. As a final aspect of the present chapter, we need to look at a striking implication of hermeneutic biology regarding the nature and status of organisms in relation to human culture.

The life of form: aesthetics, efficiency and ethics

The beekeeper's son learned from the bees that the relevant question to ask is not, 'What is the meaning of life?' but, 'What is a life of meaning?' Now we begin to see that a hermeneutic biology encourages us to ask, not about the form of life, but rather to see the living being as an expression of the life of form. The organism is the consummate craftworker who creates forms that are simultaneously effective and beautiful. Organisms satisfy their needs in extraordinarily efficient ways with bodies and behaviours that continuously inspire in us a sense of consummate beauty and grace. The craft tradition of human culture has achieved similar levels of aesthetic and efficient, effective form and function in traditions of pottery, weaving, construction of dwellings, use of the land, boat building and so on. We have tended to believe that our possession of language puts us at the pinnacle of the creative evolutionary process. However, the recognition that every single species has evolved a language with a text, the genetic thesaurus, from which meaning emerges in the process of creating the individual organism, means that we now take our place as simply another instance of this expression of living meaning. However, it becomes very clear that, compared with our biological cousins, we have become extraordinarily clumsy and destructive in the construction of our artefacts. We do not use resources and energy efficiently, as organisms do, and we often fail badly in the aesthetic quality of our artefacts. This leads us to the realization that the assumed superiority of the human in possessing language and culture while nature was simply complex mechanism, is now clearly another failure in our understanding of the evolutionary process. It is nature that not only possesses language, culture and art, but simultaneously creates its own functional beauty with efficiency and harmony.

The task before us now is to rethink our place in the stream of creative emergence on this planet in terms of the deeper understanding of the living process that is now taking form. The life of form, of which we are a part, unfolds toward patterns of beauty and efficiency that satisfy both qualitative and quantitative needs in such a way as to maintain diversity of species, cultures, languages and styles of living. Organisms provide us with the models, the touchstones, whereby we can measure our cultural achievements. Unless and until we manage to reduce our footprint on the planet to the level achieved by organisms, and simultaneously enhance the beauty of this blue planet in the way they have done, we have failed to engage with our proper destiny. Darwin shook our culture out of a complacent satisfaction with the uniqueness of our special creation by showing us that we are one with the biological realm, part of the same history of emergent creation on earth. However, we hung on to a sense of superiority over other species because of our unique evolutionary achievement: the emergence of language. Hermeneutic biology now challenges this belief by suggesting that what we express is basically what all organisms express, the ability to make relevant meaning in created form. However, our cultural style in industrial civilization fails to match the level of efficient beauty achieved by other species. Far from being the most advanced species, it turns out that in terms of living form we are failing badly, failing not only ourselves but also our co-creators of this realm. There is a new mimesis emerging in which we need to measure our performance against that of our superiors and try to mimic them. This is the message of ecological design, of biomimicry, of walking lightly on the earth and enhancing its beauty. At this point science, technology, art and ethics become one creative, indissoluble enterprise, as will be explored later in Chapter 7. The next chapter examines the life of form and the form of life in relation to principles of natural creativity.

5. The Life of Form and the Form of Life

The creativity of the natural world is expressed through form. There is no matter without form or form without matter, whether we describe the matter as dead or alive. In our scientific story about the origin of the cosmos starting with the Big Bang fireball, we imagine that initially there was nothing but undifferentiated energy at immense temperature with no form. This 'expanded' through hyperinflation to create the potential for form that we call space-time. However, for the world of form to come into being, the initial symmetry of this energetic domain of potential had to break so that some kind of self-difference emerged: some distinction arose within the uniformity, rather than the whole remaining absolutely homogeneous, which is equivalent to nothingness. How this occurred we have no idea, but every creation myth recognizes the initial emergence of form from chaos as the primal act of creation. In our Western tradition we have the Greek story of Gaia, the first divinity and the goddess of the Earth, emerging from primordial chaos. Hesiod, the Greek poet who lived in the eighth century BC, tells in his *Theogony* how Chaos, the yawning chasm of disorder, gave rise to Gaia and Eros. These three powers or beings constitute the mysterious, magical forces underlying the whole of creation, Eros being expressed through the forms of nature. They were worshipped by the Pythagoreans, among others, as the Orphic Trinity that takes its name from Orpheus the legendary musician, physician and spiritual leader who was also variously known as Dionysus, Osiris, Marduk, and Shakti in ancient traditions. And of course this Orphic Trinity of Father, Mother and Love became the basis of the Christian Trinity, though transformed by a patriarchal culture into Father, Son and Holy Ghost (see Abraham 1994, for a survey of these ideas in ancient and in modern forms).

We recognize the Pythagoreans as mathematicians and mystics who saw numbers and music as having divine creative powers. One version of their creation story has it that 'In the beginning was All-Sound,' a sea of vibrations from which emerged the world of form. This is closer to our current origin story in string theory and the Big Bang scenario than St John's 'In the beginning was the Word,' the meaningful utterances of sound that brought forth the creation. The Pythagorean teaching has been expressed as follows:

> *The seven notes of music on the physical plane represent the creative aspect of the entire mathematical principle on which the cosmos is built. When we have become capable of understanding this principle by the practice of music and the study of mathematics, our task will be to apply it to build a world of beauty and symmetry.*

The Western tradition continues to emphasize mathematical insight as a path to the Logos behind the Word, allowing us to intuit what may be lurking in the chaos as a source of intelligible form that can be described and grasped by reason. This is the direction I shall use in this chapter, asking what our scientific creation stories may be pointing to that gives us insight into the emergence of form, the expression of beauty in the natural world.

Phase transitions and the emergence of form

We witness the coming-into-being of form everyday: the formation of rain from water vapour in the clouds; icicles forming from water trickling from rocks; the emergence of leaves and flowers from the growing tips of plants. We also witness the dissolution of form, as when water turns into steam when the kettle boils, the ice melts into water, or flowers wilt and die, returning to the earth to emerge again in another form. Changes of form in matter, such as gas to liquid or liquid to solid, are known as phase transitions, one phase or state of matter turning into another. They have very distinctive properties that have been extensively studied in physics under the general term 'critical point phenomena,' which turn out to have remarkably profound and significant properties for understanding how order emerges from disorder. The insights that have been revealed in physics are now leading to a deeper understanding of the nature of life and its relation to the world of form.

Two phase transitions have dominated the physical study of critical point phenomena: the transition of a gas to a liquid and the presence of magnetic properties in a ferromagnet, such as a compass needle below a critical temperature, and its absence above this critical point. These two phenomena do not at first sight appear to have anything in common except that the transitions in each case are sudden. In the case of a gas, say water vapour condensing to the liquid form, there is a sudden decrease in the pressure and the water occupies much less volume than the gas. Liquid water also behaves quite differently from its vapour phase, exhibiting new properties such as cohesion and incompressibility and the capacity to dissolve many different substances such as salts and sugars. A ferromagnet, at a temperature below 770°C suddenly attracts iron filings and orients in a magnetic field such as that of the earth. What could these two diverse phenomena possibly tell us that is of universal significance to understanding how form comes into being?

Physics achieves its insights through a judicious combination of experimental observation and theoretical insight that comes from mathematical models of the phenomena being studied. A gas is described as a collection of molecules with a high enough kinetic energy to keep them in rapid motion. They occasionally collide but bounce off one another without sticking together. However, when the critical temperature is reached the molecules begin to stick together in clusters. At the phase transition these clusters have a size distribution that we encountered in the previous chapter. They obey a power law like that described in Chapter 1 for the sizes of ant clusters in a colony at the critical density, which is also a phase transition from chaos to order. This means that there are many small clusters, a very few very large ones, and clusters of all sizes in between that define a straight line when plotted appropriately on a graph. As the critical temperature is approached from above, clusters grow very rapidly and some extend over the full size of the container. It is as if the system is exploring all possible states available to it, including becoming a new whole that is the liquid form.

The situation regarding the appearance of a magnetic field in a ferromagnet seems to have nothing to do with the condensation of a gas. However, it turns out that these two very different states of matter behave in not just similar but identical ways. As the critical temperature is approached from above, the tiny molecular magnets of which the metal consists begin to align with one another, forming small magnetic fields whose length grows just as do the clusters of molecules in the gas. At the critical temperature these fields have a size distribution that behaves in exactly the same way as the condensing gas, following a power law of size distribution. However, the ferromagnet has an additional degree of freedom compared with a gas, which can only condense to a liquid. The magnetic field can have its north-south polarity oriented either one way or the opposite way along the metal. Which orientation develops depends upon chance, or upon external magnetic influences if there are any. But whichever polarity emerges, it does so according to the same universal principles as any phase transition, in any physical system.

These surprising discoveries of universality in critical point phenomena were the result of intensive research during the 1960s, 70s and 80s, when it gradually dawned on physicists that something almost miraculous is at work when new order comes into being. In studying the transitions from gas to liquid in substances as different as oxygen, neon and carbon monoxide, it emerged that the physical details about the different molecules involved were essentially irrelevant to their behaviour at the transition point. It required some deep mathematics to provide insight into what was going on. This was initially provided by Leo Kadanoff at the University of Chicago and then by Kenneth Wilson of Cornell University, who was awarded the Nobel Prize for his work in 1982. These two physicists developed the mathematical tools that brought power laws, scaling properties and self-similarity into sharp focus, demonstrating that at a critical point a system undergoing transition from one state of order to another looks the same at all dimensions. This is the deep meaning of self-similarity: no matter what length you choose as your unit of measurement, the system looks the same because the clusters of order are all self-similar. What is operating here is geometry, the emergence of order, which transcends physical detail. Even in the realm of relativistic quantum mechanics, where Einstein's revelations about space-time and quantum insights into the world of elementary creation come together in the form of gauge theories, based on symmetry and geometry in quantized space-time, phase transitions obeying power laws describe the processes that give rise to elementary particles, from photons to Higgs bosons, currently a focus of research. And recent experiments at the Brookhaven National Laboratory, where physicists are attempting to recreate the conditions of the Big Bang, there is good evidence that a quark-gluon plasma has emerged, again involving a phase transition of atoms back to primordial matter (Brumfiel 2004). On a totally different length scale, the coastline of Britain or Norway, or the surface of the moon, also have the property of self-similarity observed in phase transitions. This has been described in many books on fractals, the geometry that was developed by Mandelbrot to describe self-similar structures with fractional dimensions (see Per Bak 1996; Buchanan 2000). The earth itself is in a critical state, according to observations on the frequency distribution of earthquakes of different size, which also obey a power law relation. It begins to seem that wherever there is order coming into being, there are power laws as indicators of creative process. What can this mean? To find out, we need to plunge further into the quantum realm.

Quantum coherence

The description of phase transitions that occur in quantum mechanics takes us to another level of understanding of what happens when order comes into being in the so-called non-living world. A familiar example of order at the quantum mechanical level is provided by a laser, in which the photons are coherently aligned with one another regarding their phase (that is, they are oscillating as electromagnetic waves in alignment with one another). Lasers thus do not undergo the scattering of light that normal, incoherent sources like light bulbs do, so that a laser beam can travel from the earth to the moon and back without losing much of its initial order, returning as a sharp beam of light. Such a state of coherence occurs when large numbers of particles collectively cooperate in a single quantum state.

Quantum mechanics has a distinctive and precise way of describing this state of coherent order. We need to consider the relationships between the pattern of events at one point of space, say X1, and a neighbouring point, X2, Order is described over a spatial domain in terms of the behaviour in time of observable events, such as the arrival of photons recorded by instruments located at different points. The whole space-time pattern is referred to as a field, described precisely by mathematical functions of position and time in the field. When a field is coherent, there are strict relationships between events recorded at different positions in the field by what are called correlation functions. Such a function gives a measure of the frequency with which events at the two points are correlated with each other. Coherence is a distinctive and well-defined condition in which, if you look at the pattern of arrivals of photons at two points of space, X1 and X2, they will appear to be uncorrelated. However, there is a subtle relationship between them such that over an extended period the events are observed to be correlated. This is a kind of quantum entanglement such that the events appear to be independent but are in fact an expression of a single quantum state, not a mixed state.

Quantum mechanics reveals some strange, counterintuitive properties about possible forms of behaviour in the elementary events that make up our world. These describe the world coming into being and passing away all the time, with creation and annihilation operators defining the conditions under which new particles can appear and disappear in well-defined patterns of relationship. Quantum coherence tells us about conditions of order that are subtle and intriguing, reflecting the holistic nature of the quantum realm. The description above is for coherence between pairs of points in a field, but coherence can be defined for three or more points simultaneously. Here is a situation in which events occurring at any point of a field appear to be occurring independently of one another, but in the coherent state they are all correlated as defined by a correlation function.

This seems very mysterious, but there is a way of looking at this that suddenly makes more sense. Mae-Wan Ho points out that quantum coherence is not a condition of uniformity, where every component of the system must be doing the same thing all the time. She invites us to think of the situation like that in a large jazz band in which each musician is free to do his or her own thing, but all the participants together create a coherent musical whole. Freedom and coherence live together in this state:

> A quantum coherent state thus maximizes both global cohesion and also local freedom! Nature presents us with a deep riddle that compels us to accommodate seemingly polar opposites ... (Ho 1998)

The subtlety and the significance of the coherent condition begins to take on new meaning and depth in terms of this metaphor. Whenever form or order comes into being we have seen that, in physics, power laws describe the patterns of nascent order that emerge, whether in macroscopic or in quantum processes. These are states in which each component is free to participate in the formation of order on any scale as the whole explores its possibilities for expressing emergent form. Ho extends these ideas to describe the very essence of the living state:

> The organism is, in the ideal, a quantum superposition of coherent activities over all space-times, this pure coherent state being an attractor, or the state towards which the system tends to return on being perturbed.

Despite the attractiveness of quantum coherence as a condition that could begin to explain the deeper mysteries of the living condition, in particular the issue of consciousness that was explored in Chapter 3, there is no consensus on how quantum coherence can be generated and sustained over physiologically significant regions of an organism, such as the brain. The suggestions of Hameroff and Penrose implicating microtubules in such a process remains speculative, with no direct observations to support it. A volume by Globus *et al* (2004) carries this argument forward, with some interesting evidence of coherence being sustained over short length scales between organized water molecules by Fleischman, but there does seem to be a conceptual gap remaining in the applications of quantum theory to macroscopic regions of living tissue (Clarke 2005).

We saw in the last chapter that power laws describe the properties of many aspects of biological pattern and form, whether it be the connectivity of proteins in a cell, the distribution of interbeat intervals in the dynamics of a healthy heart, or the frequency of species extinction events during the course of evolution. Add to this the fractal structure of the vascular systems in plants and animals whereby sap and blood are delivered to all parts of the organism (West *et al* 1996), and the observation that in an undisturbed rainforest, gaps have a fractal distribution (Solé and Goodwin 2000), and what begins to emerge is the suspicion that in dynamic systems with power law behaviour we are looking at an indicator of creative emergence in the world. Whereas in macroscopic physics this dynamic arises only under special conditions, as at phase transitions, it looks as if organisms occupy the state of self-similarity in their dynamics and their structure most of the time, if they are healthy; that is if they are dynamically whole. Since the living realm is continually engaged in the process of generating order and dissolving it, it seems consistent that living beings should reflect this condition of emergent process in their form and behaviour. Mae-Wan Ho has suggested that the state of health in any organism is to be understood precisely in terms of maximum freedom to the components, maximum coherent order to the whole, a graphic description of the dynamic, subtle order revealed by the healthy heartbeat in Chapter 2. Whether this is a form of macroscopic quantum coherence, or a realization of a similar dynamic state in living beings, is yet to be decided.

The distinction between living and non-living now becomes rather blurred. Why should we say that a growing tree developing a fractal vascular structure is alive but a condensing gas with a fractal distribution of growing molecular clusters is dead? Both are undergoing similar processes of developing order, with similar dynamics. I am not suggesting that we abolish the distinction between these processes, because an understanding of reality requires that we discriminate between different forms with different properties. This is so that we may behave appropriately towards these different manifestations of creative order in the world, which includes making appropriate use of them for our own purposes, the goal that Francis Bacon set for humanity in developing modern science. However, a real understanding of nature goes well beyond this to learning proper behaviour in relation to other beings. We are behaving towards nature like Gingile in relation to the Honeyguide, taking all and sharing nothing. Can we find a more coherent set of principles than those that operate in Western science to bring us into a relationship of responsible participation with all the other beings of the planet, and indeed of the cosmos? In our search for the principles whereby form comes into being, we may be seeing something important about this relationship. The question I want to address is: what is the agency at work throughout the cosmos that gives rise to its continuously creative behaviour, whether in the 'dead' or the 'living' state? I am going to suggest that some form of feeling, of sentience, is involved, as have many others.

Feeling, form and meaning

If you ask a jazz musician what kind of experience is involved in participating in a really good jam session, the answer is likely to be something simple and direct, like: 'It feels really good.' The same sort of response will come if you ask someone to describe what it's like to experience health. Feeling good is not the most discriminating description of what you are experiencing in these states, but it conveys a lot. On the other hand, suppose you were asked what it feels like for photons to become coherent in a laser beam? You would no doubt say, don't be ridiculous. I can't possibly answer that because it's not part of my direct experience. However, if the laser state was described to you both precisely, and metaphorically in terms of a coherent jazz session, and you were then asked to try to imagine what it feels like, then you might be more encouraged to enter into a sympathetic relationship with the state and have a go. However, you would probably have no confidence that you were describing anything real; you were simply projecting your feelings on to the laser. A poet might do that, but certainly not a scientist! However, in order to convey to others something about their understanding of the world, scientists have to use metaphors. This is the only way we can share the meaning of our insights, however they may have been revealed. Physicists have had to resort to descriptive properties such as strangeness and charm to refer to the properties of elementary particles such as quarks. This whole language is clearly playful, expressing a real sense of the mystery of these beings and real wonder at their behaviour. The same is true of the language used by geneticists to refer to genes whose mutant forms produce bizarre morphologies in the fruit fly: hunchback, hedgehog, fringe, and lunatic fringe. As we saw in Chapter 4, science is divided into areas of research with local rules of practice that hold together a group of participants who share their understanding with one another and with the outside community in terms of agreed metaphors that convey the meaning of their insights. It is assumed that nature is self-consistent, but discovering the principles that underlie this consistency is an endless process since nature is continuously creative, always revealing unexpected phenomena that are unpredictable from our present knowledge. This is the insight of complexity, indicating that we will never have a unified theory of everything, which could describe only to a dead, uncreative universe.

So what is the agency throughout the cosmos that underlies its continuous creativity? This sounds like we might be looking for a philosophical answer, a principle that is universal. However, that would contradict
the realization that there are no universal principles that we can know for
all time in a creative universe, one based on process and change. So if
we are going to explore this question, we have to recognize that we are
really asking: what sort of description of natural creativity is appropriate to the here and now, the local situation in which we find ourselves,
that could give some guidelines about action that addresses the current
dilemmas and difficulties we face. Of course 'local' doesn't any longer
mean restricted to a place or a country; it now means being within a
community that is joined by some network of interests and procedures,
and commits itself to consensual action that is judged to be appropriate
to the present situation. Insofar as the community can decide, this is the
course of right, or skilful, action for now.

A major difficulty we are facing at the moment that separates us both
from ourselves and from the rest of nature is that of dualism. We tend to
split nature into matter and mind, our own reality into body and soul, or
equivalent dualisms, each with its own principles that are quite distinct
from one another. Descartes called matter *res extensa* and mind *res cogi-
tans,* united only in God. Since God has died, as Nietzsche announced in
the nineteenth century, they have no union. So we are left with a dualism
that ties us in a world knot, to use Schopenhauer's term for our existen-
tial Cartesian dilemma of experiencing both within us but denying their
necessary connection. There are many ways out of this difficulty, but I
believe that there is a path that is particularly appropriate now. It is the
one that I explored briefly in Chapter 4 in connection with meaning,
which Caws (1988) identifies with two more fundamental processes of
mattering, which relates to feeling, value and purpose; and signifying,
which is the mark of the intelligible. He says: 'The roots of mattering lie
in the structure of biological needs, and it is in the intelligent satisfaction of those needs that meaning first comes into play. As the immediacy
of biological need lessens in what we call advanced cultures, complex
patterns of intelligibility are pursued for their own sake and we have the
appearance of literature, music, art, and science.' This view implies a
continuity of feeling throughout the biological and cultural realms as a
primary aspect of experience. The presence of feeling in human thought
and awareness is now acknowledged by many philosophers and neuroscientists, and is certainly recognized as common sense by most people.

This is not a new position for philosophers to adopt. Suzanne Langer in a book called *Mind: An Essay on Human Feeling* (1988) has this to say: 'Feeling, in the broad sense of whatever is felt in any way, as sensory stimulus or inward tension, pain, emotion or intent, is the mark of mentality.' And Francisco Varela, with colleagues Evan Thompson and Eleanor Rosch, developed the unity of thought and feeling in their 1993 book *The Embodied Mind: Cognitive Science and Human Experience* where they discuss 'how knowledge depends on being in a world that is inseparable from our bodies, our language, and our social history — in short, from our embodiment.' In the next chapter I shall examine an earlier, intense exploration of this perspective and its implications in the writing and work of Goethe and his contemporaries during the so-called Romantic age of European thought as a model for the unity of the arts and the sciences. Here I wish to see how to make sense of our own experience of nature as revealed in the life sciences and in physics.

The great insight of evolution is recognizing the continuity of species from their earliest emergence as single-celled organisms up to the complexity of the higher mammals, including the primates and us. All the members of different species survived for some period of time, expressing the fact that their lives mattered to them as they intelligently sought to satisfy their needs. The very word 'mattering' for feeling speaks of its embodiment in material form. Why does it matter to you, we ask, appealing for an explanation in terms of feelings and values. We recognize the expression of mattering, of feeling, in the lives of others simply through their persistence and ingenuity in seeking to satisfy their needs. By extension, it is natural to assume that the lives of all organisms express the experience of feeling. We encountered in Chapter 4 the view put forward by Maturana and Varela (1987), as well as by others, that organisms have embodied knowledge of the world in which they live. Now we can add that they also have embodied feelings. The world of biological form (morphology and behaviour) reveals itself as the expression of meaning through the appropriateness of body structure and activity to context. Just as we express the meaning of a word through its contextual relation to other words in a sentence, so an organism expresses its meaning through its form and its dynamic relationships to its habitat.

However, we humans have ceased to behave intelligently in seeking to satisfy our needs. We are rapidly destroying the nature on which our survival and the quality of our lives depends. It is becoming more and more evident that this is primarily because we are still stuck in the dualist trap of splitting mind from matter, thoughts from feelings. We try to think our way out of problems instead of feeling our way, cultivating our capacity for direct or intuitive knowing as well as inferential reasoning. We have a pernicious habit of denying the body that is connected to our Christian myth of sin and redemption, expressed as the innate unworthiness of our bodies and the potential salvation of our souls. If our bodies are demeaned and our minds exalted, we will naturally demean the body of the earth and exalt the power of the intellect *(our* intellect) to sort things out through a future salvation. This is our habit of living in the future, not the present. It is a very effective way of becoming anaesthetized to present ugliness and destruction. Our capacity for denial is limitless, as explored by Derrick Jensen in his powerful book, *A Language Older Than Words* (1999). What is on offer in our culture is future security through investment in an increasing capacity to destroy others, which is the politics and economics of war and terrorism fed by fear. These are all issues that appear to be globally overpowering, insoluble, paralysing us and so making us easily governed. The politics of fear destroys our capacity for sensible action.

Feeling our way

The great mathematician/physicist/philosopher Alfred North Whitehead, who wrote with Bertrand Russell the classic volume *Principia Mathematica* dealing with the logic of scientific understanding, was deeply disturbed by the moral paralysis in our culture that he believed was connected to our mind/matter dualism. He expressed his concern as follows:

> A scientific realism based on mechanism, is conjoined with an unwavering belief in the world of men and of higher animals as being composed of self-determining organisms. This radical inconsistency at the basis of modern thought accounts for much that is half-hearted and wavering in our civilization. It would not be going too far to say that it distracts thought. It enfeebles it, by reason of the inconsistency lurking in the background.

The dilemma of course is the belief that self-determining organisms take action on their own behalf, hence have degrees of choice, but they are nothing but mechanisms, 'like an airliner only more complicated' as Richard Dawkins puts it. I believe that there is now an overwhelming biological case for recognizing the reality of feelings and intelligence in organisms, which together imply discrimination and choice. However, there remains the question how feelings evolved in organisms from the 'dead' or insentient matter of which they are believed to be composed. This issue was raised in Chapter 3, where it was concluded that there must be some quality of 'matter' that makes possible the emergence of feeling in organisms. This is another insight of complexity: emergent properties cannot come from nothing. Whitehead's work fully supports this position, which has become known as pan-sentience: matter has some quality of feeling or experience. Whitehead also argues that matter has some element of freedom. We can now try to put this together with the earlier descriptions of quantum coherence and the coming-into-being of form in nature.

The metaphor of maximum freedom to the parts, maximum order to the whole for the state of matter when organized form such as a liquid or a laser is forming, clearly suggests some element of exploration, of feeling one's way into the new possibilities. The traditional physics description of the state selected is one of minimum energy, or least action, for the conditions prevailing. It is the condition that I used in Chapter 4 to describe how the fractal structure of the self-referential molecular networks that make sense of genetic information as proto-languages emerge in functional cells during evolution. This minimal action criterion is used to describe the preferred state for many different physical systems: soap bubbles, a hanging chain, the relations between the molecules that make up the membranes of living cells, and the paths followed by photons in space-time. The time-honoured variational principles of physics provide the mathematical form for determining the minimum energy condition of any physical system. Many organisms take forms that obey this principle, as shown by the polymath, D'Arcy Wentworth Thompson in his celebrated volumes entitled *On Growth and Form* (1917, 1961). He also showed that the crystalline structure of the bones in our bodies follow this rule, as do the spirals of sea shells, the horns of ungulates and of the narwhal, the surface structure of little marine organisms called diatoms, and, of course, the arrangement of hexagonal cells in a honeycomb. The arrangement of leaves up a plant stem, called phyllotaxis, can be understood in terms of such a principle, as shown by two French physicists, Douady and Couder (1996). However, many of these forms are not fractal. It appears that forms in nature are expressions of the same principle, a generative fractal phase being followed by the deepening of the energy well of the attractor as a stable form is produced which need no longer be fractal. An extension of the structuralist principles explored in Chapter 4 has been emerging in connection with the natural geometry of biological form, consistent with these principles of doing what comes naturally by following paths of least energy or effort. An elaboration of the principles of this approach, based on a dynamic geometry of biological form that is grounded in topology (a mathematical theory of continuous transformations of forms), is presented in a book by Stuart Pivar entitled *Urform* (2007). This approach resonates with Goethe's ideas about biological form, discussed in the next chapter, which are acknowledged in this book. Together with the ideas about language and hermeneutics discussed in Chapter 4, this provides a framework for a comprehensive biological theory that links gene activity with the generation of the structure of the whole organism as a creative unfolding of meaning into living form.

It seems natural to us now to accept that a minimum energy principle describes many physical and biological forms, since we take it for granted that nature is efficient, economical. However, how do these natural processes arrive at such beautiful, elegant, symmetrical forms? How do they know when they have arrived at such a state? Are they obeying a general principle of skilful action, the appropriate behaviour for the context in which they find themselves, just as we do when we are attuned in our relationships with others? If we are going to follow the logic of emergent phenomena and accept that in order for feeling to emerge in organisms, there must be a precursor property in the matter of which they are composed, then we have to start looking again at physical principles and ask if the order we see in nature is there by a preference consistent with minimum energy or least action but arising from a quality of feeling among the components that participate in generating the order. This is rather like the experience of a dancer or an athlete who performs with most grace and beauty when achieving a condition of least effort, a condition of effortless flow and coherent movement. This is the ultimate in skilful action by whatever agency is creating a coherent whole. What we are seeking is a way to put mind and matter together through feeling or experience in such a way that they actually belong together as a unity; they are not simply stuck together by an arbitrary say-so.

Fields and souls

In science, the concept of a field is used to describe patterns of order in natural processes that are extended in space and show regularities of behaviour in time. It is worth looking at the evolution of this concept in modern science, for behind the brilliant light cast by mathematical descriptions of natural motion there lies something deeply significant in relation to our question of how to connect mind and matter in an intuitively satisfactory manner.

The first example of a field principle in physics was Newton's celebrated gravitational law, which described mathematically the universal attraction between bodies with mass, such as the planets or an apple falling to earth rather than flying off into space, as would happen if there were no such attractive force. This mysterious action at a distance without any wires or mechanical attachments between the bodies was regarded as a mystical, occult concept by the mechanical philosophers of the seventeenth and eighteenth centuries. They condemned Newton's idea as a violation of the principles of explanation in the new science. However, there is a healthy pragmatic element to scientific investigation, and Newton's equations worked too well to be discarded on philosophical grounds. Newton himself regarded the principle as an expression of the mind of God at work in the universe, though he didn't use this as a justification for his law. The gravitational field allowed him to provide an explanation for the elliptical orbits of the planets so carefully observed and measured by Kepler. This was a major triumph of scientific method, combining exact observation with mathematical description of the regularities of planetary motion.

Another celebrated example of a physical field came from the experimental work of Michael Faraday on electricity and magnetism in the nineteenth century. He talked about fields of force that extend out in space from electrically charged bodies, or from magnets. Faraday's painstaking and ingenious work described how these fields change with distance from the body in precise ways, as does the gravitational force. Again these forces were regarded as mysterious since they travel through apparently empty space, exerting interaction at a distance that cannot be understood mechanically. However, so precise were Faraday's measurements of the properties of electric and magnetic fields, and so vivid his description of the fields of force associated with them, that

James Clerk Maxwell could take his observations and put them directly into mathematical form. These are the famous wave equations of electromagnetism on which our technology for electric motors, lighting, TV, communications and innumerable other applications is based. When you have this kind of solid evidence that these fields are real, objections to their mysterious properties melt away and scientists get used to their strangeness.

The next scientific revolution connected with fields came early in the twentieth century with Einstein's reformulation of Newton's mysterious gravitational force by an even more mysterious property of space itself: it bends or curves under the influence of bodies with mass. Einstein's relativity theory did away with a force of attraction between bodies and substituted a mathematical relationship between mass and curvature of space-time. The result was a whole new way of understanding motion as natural, curved paths followed by bodies that not only cause the curvature but follow it. The universe was becoming intrinsically self-organizing. These properties were integrated with a mysterious property of light whose velocity is limited, affecting all interactions and communications throughout the cosmos. Subjects as observers made an entry into physics.

As if Einstein's relativity wasn't enough to shake up the world known to science, the next revolution was even more disturbing. Quantum mechanics, emerging in the 1920s, did away with the classical notions of fields as smooth distributions of forces through space-time and described interactions at a distance in terms of discrete little packets of energy that travel through the void in oscillating patterns described by wave functions, of which the solutions to Schroedinger's wave equation are the best known. Now we have not only action at a distance but something infinitely more disturbing: these interactions violate conventional notions of causality because they are non-local. Two particles that have been joined in an intimate relationship within an atom remain coherently correlated with one another in their properties no matter how far apart they may be after emission from the atom. Einstein could not bring himself to believe that this 'spooky' implication of quantum mechanics could possibly be real. The implied entanglement means that there is a holistic principle of connectedness in operation at the most elementary level of physical reality. Quantum fields have subverted our basic notions of causality and substituted a principle of wholeness in relationship for elementary particles.

Now I come to biology, which continued on its merry way through-out the twentieth century, seeing all interactions within and between organisms as mechanical and 'normal' despite the revolutions in phys-ics. Biology was sensible after all: genes carry information that make molecules through conventional chemical forces and these molecules make up the whole organism, following instructions written in the genes. This is complex, no doubt, but not mysterious. However, the complete reading of the book of life in DNA, the major project in biology during the last two decades of the twentieth century, did not reveal the secrets of the organism, as described in Chapter 4. It was a remarkable achieve-ment to work out the sequence of letters in the genomes of different species, human, other animals, plants, and microbes, so that many of the words of the genetic text of different species could be deciphered. Unfortunately, we were unable to make coherent sense of these words, to put them together in the way that organisms do in creating themselves during their reproduction as they develop into beings with specific mor-phologies and behaviours, the process of morphogenesis. What had been forgotten, or ignored, was that information only makes sense to an agent, someone or something with the know-how to interpret it. The meaning was missing because the genome researchers ignored the context of the genomes: the living cell within which genes are read and their products are organized. The organization that is responsible for making sense of the information in the genes, an essential and basic aspect of the living state, was taken for granted. What is the nature of this complex dynamic process that knows how to make an organism, using specific information from the genes?

The organized dynamic process that produces the form of the organ-ism is known as a morphogenetic field, which can be regarded as the pattern of relationships that exist in a developing organism at different levels of organization. This is a complex combination of activities that have biochemical, electromagnetic, informational, computational and mechanical aspects, all integrated into a self-organizing process that generates the organism as a coherent whole, organized in space and time in its morphology and behaviour.. These include the self-referential networks of molecules and their interactions described in Chapter 4 that read and make sense of the information in the genes, which themselves are part of the field as a unity. It is the field that also copies the DNA in the process of cell division and growth, which is an essential aspect of development. In Chapter 4 the networks that read and make sense of the genes were described as having properties that relate them to languages in their relational structure, with the consequence that multiple possible choices of molecular interaction coexist as ambiguities that are resolved by context (time in development, position in the embryo, external influences, and so on) into a simple, coherent differentiated state. This suggests that biology and culture are not so different after all. Both appear to be based on historical traditions and languages that are used to construct patterns of relationship embodied in organized systems, either of cells in organisms or of individuals in communities. These self-organizing activities are certainly mysterious, but not unintelligible. I and a colleague at Schumacher College, Philip Franses, with whom I have been working on these issues, see this situation as essentially similar to the situation described in quantum mechanics, where multiple possibilities also coexist and are resolved by context into an appropriate simple state. We suggest that biology has finally reached the point where the mechanical worldview fails to explain the basic phenomena of life, and a new view of ambiguity with well-defined experimental consequences (for instance, power law distributions of signalling molecules) begins to describe the principles that underlie life's intrinsic creativity and adaptability.

One further surprise from the field concept may be in store for us. Telepathy, communication between separated minds without any understood communication channel, is deeply taboo in science. However, evidence for it is accumulating and one of the most active researchers in this field is Rupert Sheldrake (1981, 1988, 1990). He uses the concept of a field of morphic resonance to describe the process whereby some kind of communication between minds can occur, irrespective of spatial or temporal separation. What the nature of this may be is as mysterious as quantum entanglement or the relativistic fields that describe dynamic behaviour and electromagnetic communication in the cosmos. Without a mathematical description of telepathic principles, it may be more difficult to get consensus on telepathy from scientists. However, scientific method puts consistent evidence and an open but critical mind on an equal footing with theoretical description, so the focus can remain on experimental results while our worldview catches up with the phenomena, if need be.

Looking at the history of ideas in modern science, it becomes evident that the field concept has acted as a kind of trickster or conjuror, taking on different forms that suit the needs of the situation observed. Goethe called the creative organizing principle in the growing tip of a plant, which gives rise to leaves and flower organs of extraordinary variety and beauty, the true Proteus, the God that never takes the same form twice, as I shall describe in Chapter 6, and his insight that all the organs of a flower are transformed leaves has been validated by twentieth century genetics. Before Newton, principles of attraction and repulsion that occur between planets, between electrically charged or magnetic bodies, were regarded as an expression of the soul of the world. Fields, it seems, function as the integrating principle of the phenomena studied in science, leading us on to ideas we thought we had left behind in the world of the Renaissance magi of the fifteenth and sixteenth centuries. Then it was love that made the world go round, as we experience in the world described by Shakespeare. With Newton, love turned into gravity, and then into many other strange and beautiful properties that speak to us of creativity with coherence and wholeness. With the convergence of nature and culture in biology we may be seeing a return of the soul of the world throughout living nature, and form as an expression of Eros, of creative relationships. Now that science has rediscovered that Logos can only be properly communicated through Mythos, the unification of science with art can proceed. And the apparently innocent field idea may help this reunion more than any other scientific principle.

6. Nature and Culture Are One, Not Two

For Leonardo da Vinci (1452–1519), the arts, sciences and technology were one indissoluble continuum of exploration and creative innovation. During the course of the Renaissance, however, there emerged a conception of art that separated it from science and technology, a separation that has remained firmly embedded in our culture despite attempts to reunite these activities in areas such as design, craft, social art, and architecture. Art in its modern conception has become the domain of subjective creativity, the exploration by the individual of forms of self-expression that reflect current trends of cultural experience and thought. The hallmark of modern art is precisely the distinct signature of the individual so that we can recognize the work of the person through the intrinsic qualities of self that are revealed.

In sharp contrast to art as the expression of personal vision, science has come to be understood as the objective description of the world by a community of scientists who dedicate themselves to overcoming their subjective bias so that they can describe reality in its universal features, true for any observer. This stark polarity between different ways of knowing and doing is based on stereotypes that are rarely realized by actual people, who are a mixture of tendencies and inclinations. However, the distinction between the artist and the scientist remains deeply embedded in our educational systems and the careers that people are encouraged to pursue. If this polarity is beginning to soften and erode, it is not the institutions that embody these extremes that are changing; it is individuals who find the stress of being exclusively one or the other intolerable and occupy a middle ground that combines the two. However we do not have a vision of how to live a whole life that is expressed in education and is catered for by career paths.

There have been significant moments in Western history since the Renaissance when the polarization of the arts and the sciences was questioned and challenged. One of these has come to be known as the Romantic Movement of the late eighteenth and early nineteenth centuries. This is often seen as a reaction to the severe rationalism of the Enlightenment, which took the view that what can be known by reason is necessarily objective, existing independently of the observer. If you want to know about subjective states, you have to look at them 'objectively,' possibly by introspection but more satisfactorily by observing them in another person. This is the scientific stance of many people currently engaged in the study of consciousness, for example, and certainly of most neuroscientists. The dilemma that this presents is the following: we know that we have feelings such as pain and joy and the experience of beauty or ugliness, and these feelings belong to me as a subject. No amount of 'objective' description can substitute for these subjective experiences, which are real. How do such experiences come to exist, and reliably relate to 'objective' states such as a wound that gives pain or a perfume that gives pleasure, if reality is entirely objective? This problem was discussed in Chapter 2, with the conclusion that subjective experience is as real as objective phenomena. The Romantics drew this conclusion also, based on the evidence of their own experience, which told them that the Enlightenment view of nature and humanity was deficient. The challenge then was to account for both subjective experience and rational reason in humanity. What is the ground of humanity? What is our real nature that we can both feel as subjects and think clearly about objects? In the following account of Goethe and the romantic movement I have drawn extensively from Robert Richards' book *The Romantic Conception of Life: Science and Philosophy in the Age of Goethe* (2002), whose primary title I have borrowed.

The Romantic conception of life

It was the German Romantics who first struggled with this dilemma, starting in the period of German culture known as *Sturm und Drang,* from about 1770 to 1800, then becoming the Romantic Movement lasting roughly from1800 to 1820. A similar movement occurred during much the same period in England when poets such as William Blake, then Wordsworth, Keats, Shelley and Coleridge, articulated their view of human nature that combined feeling and thought in the form of expressive action that characterizes a human life. A key figure in the European engagement with these issues was Johann Wolfgang von Goethe, who himself contributed to the Romantic Movement with his celebrated novel *The Sorrows of Young Werther* (1774). However, he believed that he had moved through this to a more balanced position in his other writings, and particularly in his scientific studies. Although he is generally regarded as Germany's foremost man of letters, a German Shakespeare, his own view was that his scientific work was more significant than any of his literary achievements. Since the latter remain amongst the highest expressions of dramatic literature, one wonders where to rank his scientific studies. Judgement on this has been slowly changing for a century or more, but it is only now that we can begin to appreciate what his studies of natural phenomena represent, and their promise of a more holistic approach to understanding ourselves and the rest of nature.

Goethe was born in Frankfurt on August 28, 1749, 'as the clock struck twelve noon,' he tells us in one of his writings. He therefore arrived on the European scene when science had established its dominance as the only reliable way of knowing natural phenomena. Literature and the arts were important as subjective expressions of human experience, but not as revelations of objective reality. Goethe's father was relatively well-off, with a large house containing an extensive library and works of art by local artists. Goethe's early education was at home, at the hands of his father and several tutors, and he became fluent in Italian, French and English as well as his native German. He was deeply influenced by the new poetry, which expressed powerful emotions through unconventional meter and free form. His father, however, disapproved of this literary movement and insisted that he pursue a practical career in law at the university where he had studied, Leipzig. There Goethe was deeply bored by the lectures he attended on law, legal history, and legal institutions, and continued to write poetry which he submitted to his teachers for comment. They were uniformly discouraging. Goethe characteristically expressed his disappointment and loneliness through poetry itself, often with a healthy measure of sardonic humour.

It was not long, however, before Goethe fell in with a group of sympathetic companions who appreciated his talents. One of them, a tutor in Leipzig, had a selection of Goethe's poems beautifully copied and bound, greatly encouraging Goethe in his writing. The companions met regularly at an inn where Goethe met the innkeeper's young daughter, Anna Katharina, called Käthchen. He became enamoured of her with all the passion of his temperament, straining the relationship with unfounded jealousies. This intensity of feeling was characteristic of Goethe's relations with women and would repeat itself many times throughout his life. However, he was always able to express his feelings through poetry and writing with an intensity that matched his experience, to some extent freeing himself from disturbing emotions by putting them into words:

> *Bin ich in Lieb zu ihr versunken*
> *Als ich von ihren Blut getrunken.*
>
> (I have fallen so in love with her
> It's as if I had drunk her blood.)

Goethe's poetry has such a perfect match of words to intense feelings in his mother tongue that it inevitably loses much in translation.

In 1768, after three years at Leipzig, Goethe experienced a crisis that culminated in a physical collapse and he returned home to Frankfurt where he took a year and a half to recover. During his convalescence he was encouraged by his physician to read various alchemical treatises, such as those of Paracelsus and von Helmont, as well as more conventional works on chemistry. He was also guided by Susanna von Klettenberg, an adherent of the Moravian Brethren and a friend of his mother. The combination of disease, mysticism and alchemy deeply influenced Goethe, leading him away from orthodox religion and towards neo-Platonism and the occult. However, he had his law degree to complete and, after his recovery, he went to Strasbourg, where he met a man who had a deep and abiding intellectual and emotional influence on him. This was Johann Gottfried Herder (1744–1803), an ordained clergyman who had studied with the great philosopher of the time, Immanuel Kant (1724–1804). Kant's philosophy was and remains a powerful reflection on scientific knowledge and its relation to aesthetic and moral judgement. He accepted the science of Newton as firmly established and was influenced by the sceptical philosophy of Hume, the first showing how reliable, objective knowledge of nature could be obtained, the second maintaining that we can never understand nature in its real essence. Kant argued that all we can know about nature is in the form of our ideas about it; we can never know nature directly through experience by means of direct, non-inferential, or intuitive knowledge. His idealism took the form of believing that we do not have such a capacity for directly knowing the world and we have to proceed by logical, rational inference from evidence, as in science.

Goethe initially accepted this metaphysical position, but it sat uneasily with the intensity of his experience of nature and of relationships, which made him believe that there can be an immediate, deep connection between a person and whatever absorbs them that gives reliable knowledge of the other. Herder encouraged Goethe in this approach to nature by his fascination with the unique particularity through which natural forms express themselves. The thoughtful clergyman was sceptical of the Enlightenment preoccupation with abstraction and universalism, and encouraged Goethe's instinct to immerse himself in the process of natural creation. However, Goethe's destiny at that period of his life was to continue the practice of law and to pursue his literary and amorous interests. One of these relationships elicited from Goethe the novel that was to set Europe on romantic fire: *Die Leiden des Jungen Werther* (The Sorrows of Young Werther) published in 1774. Heavily autobiographical, this short novel in the form of a series of letters tells the tale of a young man who falls in love with a woman, Charlotte, who is engaged to be married to his best friend, mirroring Goethe's own experience in Wetzlar, where he went to practise law after finishing in Strasbourg. In the novel, Werther writes to a friend: 'Never before have I been happier. Never before has my sensitivity to nature — yea, even to the rocks and grass under my feet — been richer or deeper.' Werther is, however, unable to resolve the dilemma of his passion for Charlotte and his respect for his friend, her fiancé, becoming more and more despondent. He gradually loses his sensitivity to nature, a life-giving bond, thus losing his will to live and finally committing suicide. Here is how this is described in the novel:

If you could see me now, dear Charlotte, in the whirl of
dissipation — how my mind dries up and my heart is never
really full! Not one single moment of happiness: nothing!
Nothing touches me. I stand before a puppet show and see
the little puppets move, and I ask myself whether it isn't an
optical illusion. I am amused by these puppets, or rather, I
am myself one of them; I sometimes grasp my neighbour's
wooden hand, and withdraw with a shudder. In the evening
I resolve to enjoy the next morning's sunrise, but I remain
in bed; during the day I promise myself a walk by moon-
light, but I stay at home. I don't know why I get up nor why
I go to sleep.

The leaven which animated my life is gone; the charm
which cheered me in the gloom of night, and aroused me
from my morning slumbers, is no longer with me.

Here we see the intimate relationship between the artist, nature and
the significant woman that constitutes a continuing theme in all Goethe's
work. The novel had a powerful impact on European culture, being a work
of artistic audacity that transformed the conventions of the novel from
rather sentimental, moralistic epistles to letters with a lyrical intensity previously unknown in narrative prose. Johann Merck, a poet, literary critic
and friend of Goethe, described the quality of his writing as follows:

The inner feeling that all his compositions exhibit, the living presence that accompanies the art of his representation,
the felt details that reside in all the parts of the work and
their particular selection and ordering — all of this shows a
universal and powerful master of his material.

Napoleon himself told Goethe that he had read the book several times
when they met in 1808.

Werther established Goethe's reputation as a major figure on the
European literary scene. He had returned to his father's house in
Frankfurt, but an invitation from a young prince, Carl August, in 1775
to join him in his duchy of Saxe-Weimar-Eisenach drew Goethe into a
life of service that lasted the rest of his life. In Weimar, Goethe became
a government officer, where his abilities were rapidly recognized by
advancement to the presidency of the ducal chamber. Life at court was a

mixture of government and social activities, and it was not long before Goethe became deeply involved with Charlotte von Stein, wife of a wealthy landowner and master of the stables in the duchy. Goethe was twenty-six when he met Charlotte, who was thirty-three and mother of three boys, though she had borne seven, four girls having died. Her physician wrote this account of her:

> She has large black eyes of the greatest beauty. Her voice is soft and low. Every man, at first glance, notices in her face earnestness, tenderness, sweetness, sympathy, and deeply rooted, exquisite sensitivity. Her courtly manners, which she has perfected, have been ennobled with a very fine simplicity. She is rather pious, which moves her soul with a quiet enthusiasm. From her easy, Zephyr-like movements and her theatrical accomplishment in artful dancing, you would not suspect — something that is certainly true — that a hushed moonlight and a quiet evening fill her heart with the peace of God.

When Charlotte heard that Goethe, the famous author of *Werther*, might be coming to live in Weimar, she asked a friend about him. The friend reported that he had heard from a 'woman of the world' that 'Goethe is the most handsome, liveliest, most original, fieriest, stormiest, softest, most seductive, and for the heart of a woman, the most dangerous man she had ever seen in her life.' The match was perfect, destiny presenting Goethe once again with passionate conflict. Charlotte inevitably fell under Goethe's spell, as he did under hers, and there began a relationship that elicited some of Goethe's most sublime poetry. However, Charlotte preserved her piety while maintaining a relationship that took Goethe to new heights of creativity.

Characteristically, Goethe's love for Charlotte von Stein had the effect of transforming his love for nature, where he saw Eros expressed everywhere through natural form. His duties took him on travels throughout the duchy to examine roads and to visit mines, for which he was responsible. This led to geological studies, which inspired Goethe to inquire into the origins of primitive rocks like granite and basalt, solidified from lava flows during ancient times. Thus began the more scientific aspects of his career, engaging him in practical, observational work that gave him a different perspective on nature. His interests in biology were initially centred on human anatomy, for he wanted to deepen and perfect his artistic rendering of the human form. However, this initial focus rapidly expanded to the study of comparative anatomy and questions regarding the anatomical relations between humans and other vertebrates. Goethe's intellectual acquaintance from his Strasbourg days, Herder, had come to Weimar at Goethe's invitation to join the court circle and was writing a book on human history. Herder believed that there was a unity of plan or type underlying the diversity of biological forms such that 'generally a certain uniformity of structure and a principal form seem more or less to govern, a form that mutates into multiple varieties.' He thought that the similarity of structure in land animals was fairly obvious, but there were those who claimed that the human skeleton was distinctly different due to the absence of particular features that are present in lower forms. Goethe made observations on a human embryo where he identified a bone that was believed to be absent, the intermaxillary bone, then identifying its faint presence in an adult skull. Herder used this evidence in his book. The foundations for Goethe's views on the unity of underlying form despite radical transformations in the biological realm thus grew out of sound observational evidence, views that he would later develop much more extensively in his botanical studies. Clearly the biological ideas of his time in Europe were tending towards an historical, evolutionary understanding of life on earth. Charlotte found these ideas dangerous, 'implying that humans were probably first plants and animals,' but fascinating when presented by the man she loved. 'Goethe now messes around thoughtfully in these things. And anything that first has passed through his imagination becomes extremely interesting,' she wrote.

Although professional scientists did concede Goethe's contribution to anatomical studies, there was a tendency to defer to him politely without taking him too seriously. A kind of patronizing dismissal was characteristic of the attitude towards a man of letters who dabbled in others' territory. Professional specialization had become the mark of the academic, and any claims of an amateur to make significant contributions to scientific understanding tended to be dismissed without close examination of the evidence. This became particularly true of Goethe's studies in optics and colour, where he challenged Newton's procedure and theory. He was made to suffer the consequences, being regarded as a creative genius in his own field, but who failed to understand how to conduct proper scientific experiments and construct correct theories. This continues to be the attitude today, but a shift is taking place that locates Goethe's scientific approach within an altered context of quantities together with qualities, analysis with intuition. Science evolves, like everything else.

Goethe's Italian Journey: revelations of plant form

In the 1780s Goethe developed a taste for horticulture, laying out a garden at his house in Weimar. Characteristically, he began to speculate on the nature of plant form as an unfolding of transformations based on a common form or archetype. He spent much time discussing this and the Linnean system of classifications of plants with a retired professor from Göttingen who had come to live in Jena, near Weimar. Goethe made close observations of plant growth and development, coming to the conclusion that all the organs of plants, including those of the flower, can be understood as modifications of the leaf: 'the leaf in its most transcendental sense,' as he put it. Although this work absorbed him, by the summer of 1786 Goethe felt depressed and exhausted by his constant round of duties at Weimar and by the enervating effect of his unfulfilled relationship with Charlotte von Stein. Goethe felt that, in his thirty-seventh year, he desperately needed to escape and rejuvenate his soul. And the place that drew him was Italy. One of Goethe's most famous poems, 'Mignon,' expresses this longing:

> Do you know the land where the lemon trees flower,
> Where in verdant groves the golden oranges tower?
> There a softer breeze from the deep blue heaven blows,
> The myrtle still and the lovely bay in repose.
> Do you know it?
> There! There!
> Would I go with you, O my master fair.

Goethe secretly planned his escape from Karlsbad, the summer location of the court. After the celebration of his birthday on August 28, 1786, he slipped away by coach that took him south, across the Alps and into a new world from which he would return transformed. He travelled to Verona, Venice, and Florence, staying in each city briefly before making his way to Rome, which became his home for four months. There he fell in with a group of German artists who were initially in great awe of the famous poet. However, they soon became close friends, and with one of them, the painter Angelika Kauffman whom Herder described as 'perhaps the most cultivated woman in Europe,' Goethe formed a lasting relationship. Another painter, Tischbein, who left memorable portraits of Goethe, recalls their first meeting in Rome. Goethe had been warming himself by the fire in a inn near St Peter's, when he got up and introduced himself: 'I am Goethe.' It was Tischbein's extraordinary skill in and understanding of painting that gave Goethe deeper insights into the medium and a recognition of his own modest talents. And it was Tischbein who captured the striking allure of Emma Lyon, the mistress of the English ambassador in Naples. She entertained her guests with poses in exotic costumes, totally captivating Goethe, as she was later to do with Horatio Nelson, resulting in their scandalous affair.

Goethe returned to Weimar in 1787 from his Italian experience a new man, expressed through a new beginning in his life and work. His experience of nature in Italy, particularly the exuberant, luxurious forms of plants with which he was familiar from their German counterparts, revealed to him the prodigious power of plant form to express itself in ways appropriate to context, emphasizing transformational powers. In a letter to Charlotte von Stein he wrote:

> Tell Herder that I am very near to the secret of the genera-
> tion of plants and their organization and that it is the simplest
> thing conceivable. Under these skies, one can make the most
> beautiful observations. Tell him that I have very clearly and
> doubtlessly uncovered the principal point where the germ is
> located and that I am in sight, on the whole, of everything
> else and that only a few points must yet be determined. The
> *Urpflanze* will be the most wonderful creation on the earth;
> nature herself will envy me. With this model and its key, one
> can, as a consequence, discover an infinity of plants — that
> is, even those that do not yet exist, because they could exist.
> It will not be some sketchy or fictive shadow or appearance,
> but will have an inner truth and necessity. The same law will
> be applicable to all other living beings.

This revelation was Goethe's highly dynamic conception of the archetype of the plant, on which he had been working before his Italian journey. His notion of leaf as the basic organ of the plant emphasized its creative potential, which he described as the 'true Proteus' underlying plant form, Proteus being the God that never takes the same form twice. The apparent contradiction between the notion of an ideal, Platonic archetype that allows us to recognize the essence of a plant despite its expression in many different species, and the recognition of transformational potential in living nature as the basis of evolution, is one that many biologists have struggled with, and Goethe's solution was not generally understood. He has been characterized both as a member of the idealist tradition in morphology, and as a visionary who anticipated and prepared the ground for Darwin's historical conception of evolution. However, Goethe was working towards a conception that is neither strictly idealist nor historical, but what we would now call phenomenological: nature leads us to a correct conception directly if we pay close attention to the phenomena. However, it takes much practice by the individual to develop this capacity, which Goethe regarded as a scientific method that could be developed in the future.

> There is a delicate empiricism which makes itself utterly identical with the object, thereby becoming true theory. But this enhancement of our mental powers belongs to a highly evolved age.

It is only in the past ten years or so that books describing the original nature of Goethe's scientific approach have begun to appear, acknowledging the realization that Goethe was correct in his judgement that his scientific studies were more important than anything else he achieved in his life. Among these books is a superb piece of exposition of Goethe's science and philosophy by Henri Bortoft entitled: *The Wholeness of Nature: Goethe's Way toward a Science of Conscious Participation in Nature* (1996). Scientists themselves are realizing the depth of Goethe's approach to natural phenomena, as evidenced by the acknowledgements of both physicists and plant geneticists that his observations, and particularly his method, have validity and provide us with a new way of integrating different ways of knowing. In the science magazine *Nature,* two plant biologists say in a 2001 review article:

> What controls the difference between a plant's floral organs and its leaves? Over two hundred years ago Johann Wolfgang von Goethe proposed that the different parts of

a plant result from 'metamorphosis' (meaning transformation) of a basic organ, the 'ideal leaf.' But if floral organs
are just modified leaves, what are the modifiers? *(Nature,*
409, pp.469–471, 2001)

The answer they give in the article is of course the action of specific
genes, whose influence Goethe identified as 'different qualities of sap,'
as he knew nothing about genes. Such is the power of phenomenological
observation.

A *theory of colour*

One of Goethe's goals in Italy was to learn about the theory that lay
behind the use of colour by Italian painters. However, he discovered that
there was no systematic understanding of colour, its meaning in nature
and its use in the arts. There were simply conventions and rules of thumb,
useful but not illuminating. After his return to Weimar, Goethe turned
with characteristic absorption to the study of colour when he managed to
borrow a prism from a physicist in Jena and finally found time to examine
what it could tell him about the origin and nature of colour in the natural
world. He knew about Newton's theory of colour, but as with most of his
contemporaries, he misunderstood it, thinking that Newton had demon-
strated that any source of white light could be separated into the seven
colors of the rainbow by passing it through a prism. When Goethe finally
took the borrowed prism out of his desk drawer and looked through it at
various things in his room, he was astonished, remarking that Newton
was clearly mistaken in his theory. For when Goethe looked at the white
wall through the prism, he saw no colours at all. He saw colours only
where there was a contrast between light and dark, as at the edges of the
window or the leaded panes of glass. And there he saw a colour that is not
included in the Newtonian spectrum. Suddenly Goethe became utterly
immersed in exploring an area that he believed had been fully explained
by Newton, but now presented him with many questions and puzzles.
So began a study that lasted some seventeen years, from 1790 to 1807,
ending with a book, the *Farbenlehre,* that has been pretty thoroughly
misunderstood by scientists and historians alike.

Goethe's approach to scientific work was to explore every aspect
of the phenomena which he sought to understand in order to allow the

observations to reveal their nature and origin to the investigator. This was his 'delicate empiricism.' With his prism he examined the full range of possibilities available to him, making observations on the many ways in which light and dark interact with the generation of colour when seen through a prism. He soon realized the error of his assumption about white light consisting of all colours. The experiment performed by Newton was to look at light passing through a small hole in a screen, so that the light was surrounded by dark. Then the normal spectrum of colours emerged after the light passed through a prism. However, Goethe observed that this spectrum was produced only under the rather special conditions of light surrounded by dark. At an edge where there is dark above and light below, as at the upper edge of a window, only the blue range of colours is observed (conventionally, blue, indigo, and violet); while with light above and dark below, only the red range arises (yellow, orange, red). Under these conditions there is no green. To see the normal spectrum with green between the 'cool' and the 'hot' colours requires that the light is bounded by dark. This can occur either with a small hole in a dark screen through which light passes (Newton's experiment); or by having a white card covered by two black cards which are slightly separated with the white card visible between them (Goethe's experiment). With a wide separation, looking through a prism reveals the blue colours above and the reds below, but no green. As the slit is narrowed by pushing the black cards together, suddenly green emerges. It requires special conditions for green to appear, and Newton had chosen these conditions for his experiments.

There is a complementary experiment to this: place two pieces of white card on a black card, leaving a black instead of a white slit. If this is wide, the prism reveals the red colours at the top boundary and the blues at the bottom, reversed from the previous experiment since light and dark are reversed. Now gradually close the separation, making the black strip narrower. At a certain point, a new colour suddenly emerges between violet and red: a bright pink that is usually described as magenta. This is an experiment that Newton did not carry out, yet it is part of the phenomenology of colour emergence that requires understanding. Goethe carried out these and many other studies of light, seeking always to let the phenomena reveal the generative sources of colour. What he discovered was that the physical colours observed through a prism always arise from particular relationships between light and dark. White light cannot be said to be made up of all the colours pre-existing and ready to be separated by a prism, which is the general understanding of Newton's theory. Colour arises only from the interaction of light and dark. So colour arises from relationship; it does not pre-exist as something given, such as a wavelength of light. Goethe recognized the relationship not only between light and dark, but also with the observer who participates in the experience. And he sought principles of explanation that acknowledged this. So absorbed was he in these studies that the turmoil following the French Revolution, which swept through Weimar with Napoleon and his armies causing deep disturbance to everyone, did not prevent Goethe from continuing with this work as he moved from place to place, carrying his notebooks and prism with him and performing experiments under very difficult circumstances. Goethe's revolutionary contributions to literature and science were of course part of this general transformation of political and social life in Europe, and he recognized this. Nevertheless, dedicated work requires some stability which Goethe struggled to maintain during this period in the 1790s.

A significant objective of Goethe's optical studies was to understand the emergence of colour in nature. He recognized that the red range of colours, from pale yellow to blood red, are familiar from the changes in the colour of the sun from noon to sunset; while the blue range, from light to dark, is seen in the colour of the sky, which darkens as one climbs higher in altitude in the mountains. He did not of course have the testimony of astronauts who witness the sky darkening through the blues to indigo and violet, then to black as they climb out of the earth's atmosphere. Goethe's description of these processes was to ascribe the reds to a darkening of the light, while the blues are a lightening of the dark. The feeling qualities that we experience of colour reflect this: the 'cool' blues tend to be described as enlightening, peaceful, expansive, while the 'hot' reds have the qualities of vitality, intensity, and passion. Here we encounter the value of qualitative descriptors of experience, which reflect real influences of colours as used in colour therapy or indeed in painting, the initiator of Goethe's quest to understand colour. In a deeply perceptive phrase, Goethe described colour as an expression of the deeds and sufferings of light, recognizing the subjective, participatory aspect of natural phenomena as described in Chapter 5.

The scientific recognition of Goethe's work on colour has been slow to emerge, as in the case of his plant studies. However, in a recent article in the journal *Physics Today* (July 2002) we find the following: '... chaos theorist Mitchell Feigenbaum consulted Goethe's work and was surprised to find that "Goethe had actually performed an extraordinary set of experiments in his investigation of colours".' The authors go on to remark:

> We agree with Feigenbaum that the experiments contained in *Theory of Colour* are what gives Goethe's work its abiding interest. In this article, we suggest that Goethe was a remarkable representative of a research style we call exploratory experimentation. Long ignored by historians and philosophers of science, exploratory experimentation has nevertheless played a crucial role in the history of physics. Among others, Michael Faraday's investigations into electromagnetism followed the exploratory approach. ... we tell the story of exploratory experimentation by looking at two investigations of colour from different historical periods — Goethe's experiments with prismatic colours and Edwin Land's experiments on colour vision. *[Author. Land was the twentieth century pioneer who clarified both the limitations of Newton's theory and the basic validity of Goethe's approach.]* The 'retinex' theory of colour vision that Land developed on the basis of his experiments has two essential elements: it recognizes lightness (that is, reflectance) as the fundamental stimulus of colour, and it emphasizes the importance of boundaries, which allow the eye to estimate lightness by seeking out singularities in the ratio of energy flux from closely spaced points. The parallel with Goethe's theory, which itself emphasizes the crucial roles of lightness and boundaries, is striking. (N. Reibe, and F. Steinle 2002)

Spinoza and scientia intuitiva

Before his Italian journey, Goethe had accepted Kant's view that the ultimate realities of nature are hidden in a noumenal world, inaccessible to direct knowing by humans. However, Kant had also written on the importance of the imagination in science and art in his *Critique of Judgement,* where he reflected on the expression of form in living nature. Here is a passage about this from my book *How the Leopard Changed Its Spots:*

Kant was so struck by the complex and subtle coherence of organisms that he likened the developmental process, the transformation of a simple initial form such as a fertilized egg into the adult form, to the creation of a work of art, which also has an inner coherence expressed in the dynamic unity of its emergent parts. The beauty that we see in organisms he likened to that which comes from the experience of a poem, a painting, or a piece of music. Kant saw this as the world of form, whose enjoyment depends on a free play of the mind, which means not attempting to fix the form in a category but experiencing its coherent wholeness as something of value in itself. 'The content here appears in that qualitative perfection which requires no external completion, no ground or goal lying outside itself, and it brooks no such addition. The aesthetic consciousness possesses in itself that form of concrete realization through which, wholly abandoned to its temporary passivity, it grasps in this fleeting passivity a factor of purely timeless meaning.' An organism or a work of art expresses a nature and a quality that has intrinsic value and meaning, with no purpose other than its own self-expression. Kant describes this as 'purposiveness without purpose,' using the eighteenth century notion of purposiveness as 'individual creation which displays a unified form in itself and in its structure ... A purposive creation has its centre of gravity in itself; one that is goal-oriented has its centre of gravity external to itself; the worth of the one resides in its being, that of the other in its results.' (Goodwin 1994)

Goethe read Kant's *Critique of Judgement* in 1790, and it had a profound influence on his view of the nature of organisms and how we may know them. Shortly after he began to read the work of Baruch de Spinoza, a rather reclusive Jewish philosopher who had lived in the Netherlands from 1632 to1677. Encouraged by Herder to do so, Goethe discovered a worldview that resonated with Kant's ethical and aesthetic views of nature, but differed from Kant's belief that we cannot know the world directly. Spinoza wrote in a style adopted from Euclid's *Geometry,* as a series of logical propositions that formed a coherent whole, or so he intended. His position can be summarized as follows:

1. God and nature are one.
2. Individual objects have both a material and an ideational or spiritual side.
3. One can grasp the spiritual side of things, their essential idea or archetype, in an intuition.
4. Individual parts of an object, or of nature, can be understood only in relation to the whole of which they are parts.
5. Imagination can easily mislead the careful scientist.

Here is a monism that presents nature and God as simultaneously knowable, though cautioning about the dangers of undisciplined use of the imagination, echoing Francis Bacon's warnings about human weaknesses. However, whereas Bacon's views had led to specialist disciplines within which scientists kept God well away from their work, Spinoza believed that a disciplined imagination is the route to direct knowledge of the essence of things, through a cultivated intuition. Thus pathways could be found to hidden aspects of nature so that discoveries in the soul can reveal natural truths. This was Spinoza's *scientia intuitiva*. Goethe took this as the challenge he faced in his scientific studies, seeking to link them to his creative work in the arts and both to the understanding of nature. The continuing significance of this task is evident from the recent appearance of a book *Looking for Spinoza* (2003) by Antonio Damasio, a neuroscientist who is seeking to rejoin human thought with feelings and emotions in a unified account of brain function and physiology.

Spinoza's philosophy attempts to unite not only God and nature, but morals and knowledge in a kind of naturalistic ethic. His ethics is not based on prescription, saying what we ought to do, but rather on clear understanding of the situation in which one finds oneself. Andrew Collier, in his book *Critical Realism* (1994) that is concerned with emergent phenomena in the social and political realm, understands Spinoza's position to be that a free person who is guided by reason will tend to act in the right way. In this he joins psychology to ontology, to what is, by recognizing that:

1. an emotion can only be overcome by another emotion;
2. emotions are not just data, but involve beliefs which may be more or less adequate, and the emotion consequently more or less rational;
3. we are free to the extent that we have rational emotions, based on adequate ideas.

Here is a philosophy of freedom and rational, ethical action in which a person is always in the process of transformation to better alignment with his or her context through a deeper understanding of oneself and the other, both of which can be known directly through a cultivated intuition. The challenge was to actually develop a scientific method of pursuing this, which Goethe felt drawn to explore.

A conversation with a tree

There are many different ways in which our relationship with the natural world can be experienced, but one which I shall never forget happened on Dartmoor one beautiful September morning in 2003. The group of students at Schumacher College taking part in the Holistic Science MSc that year, together with Stephan Harding and I, the residential teachers on the course, were on a field trip that involved camping overnight beside the stream that becomes the East Dart River. We were well up on the moor, with sheep and Dartmoor ponies as companions, and the group decided to spend that glorious Saturday morning hiking further up to experience the beauty of the land in the warm September sunshine. I had been reading Stephanie Kaza's book *Conversations With Trees,* so I said that I would stay behind and converse with a willow tree that grew beside the stream near our camp. I settled down, listening to the babble of the stream and simply becoming present to the tree and its context. As I slipped into deeper awareness of the tree and its surroundings, I had an experience of deepening sorrow. I realized that in the past the tree would have had many companions, willows and other species that populated the moor with dense forest cover, and it was very lonely. Humans had removed nearly all the trees from the moor as agriculture and charcoal production developed, so that by about four thousand years ago there were only a few survivors on the moor and patches of woodland in the valleys. This is our heritage, which is often seen as the natural state of this land. The tree reminded me that this is not so, and it felt isolated and lonely. As I gazed disconsolately and regretfully at the tree, realizing the enormity of our past actions, I became aware of the extraordinary beauty of what remained, and its capacity to survive and regenerate. Although what I saw was a residue of something that had been grand and magnificent, it still presented a beauty of form through transformation that was exquisite in itself. Eros was present there, with Gaia and Chaos, the Orphic Trinity continuing to express the life of the cosmos in this little corner of England.

We may celebrate and enhance this creativity by appropriate participation, or we may exploit and damage it; but we are minor players in the greater scheme of things, which will continue with or without us.

Romantic realism

Goethe had another companion on his intellectual journey to understand living form, the poet Friedrich Schiller, whom he met in 1794. These two geniuses, who became close friends, had much in common but differed sufficiently to spur each other to clarify and sharpen their views on humanity and nature. While Schiller emphasized the creative freedom of the artist, Goethe insisted on the rights and constraints arising from nature. Both men recognized the deep relationships between humans and other species, and it was Schiller who recognized the reality of play in animals, a subject that still raises difficulties for functional Darwinians who interpret all animal behaviour in terms of survival. Schiller proposed that: 'An animal may be said to be at work when the stimulus to activity is some lack, and it may be said to be at play when the stimulus is sheer plenitude of vitality.'(See also Burkhart 2005.) This view resonates with the perspective developed in Chapter 4 that puts meaning back into evolution and joins nature to culture. However, there were also deep differences between the two Germans: Schiller was a Kantian idealist, Goethe a Spinoza-inspired realist and initiator of a view that we could call Romantic realism in recognition of life as form, the expression of Eros. Their deep friendship and respect for each other resulted in dialogues which drew both toward common ground. Goethe had directed his observations on natural form to plants, using the same concept of a common type that he had learned in relation to animal form to understand the coherence and wholeness that underlies individual plants of different species despite the radical transformations that they undergo. Schiller encouraged Goethe to understand this in terms of an ideal form that could not be seen with the physical eye but only with the eye of abstract thought. However, Goethe experienced his *Urpflanze* directly as something he could actually see behind the diversity of plant form. It spoke to him as true theory revealed in the phenomena, not lying behind them as a Platonic ideal. He had the same experience of an archetype lying within the diversity of expression in the forms of animals, a view he shared with the great French comparative anatomist Geoffroy St Hilaire who saw the world of animals united under mutual transformation of their forms.

Goethean science: the methodology of delicate empiricism

Goethe did not himself describe a systematic method for the holistic study of phenomena, though it is implicit in his scientific studies and writings. These have been closely examined by a number of individuals whose interest has been to develop a scientific method that includes qualities as well as quantities, allowing full expression of the human capacity to know the world both analytically and through the cultivated use of the intuition. The resulting procedures are known as Goethean science. Rudolf Steiner had been given the task of publishing Goethe's scientific works, which appeared in German in four volumes between 1883 and 1897. A book entitled *Goethean Science* contains his interpretation of Goethe's scientific principles, translated into English and published in 1988. The movement established by Steiner, known as anthroposophy, incorporates many of Goethe's insights into educational and devotional practices that seek to integrate the different aspects of human aspirations into an integrated scientific/artistic/religious framework of understanding and action in community. A number of members of this movement have described Goethe's vision, in particular Jochen Bockemuhl (1986, 1992) and Margaret Colquhoun, in collaboration with Axel Ewald (1996). An excellent collection of essays that cover diverse aspects and examples of Goethean science is *Goethe's Way of Science: A Phenomenology of Nature* (D. Seamon and A. Zajonc, eds., 1998).

An outline of Goethe's methodology

Goethe's approach to nature is based on the way we get to know each other, nature being treated as much as a subject as an object. Hence the procedure has a foundation of common sense, turned into a systematic process that can be varied in detail and emphasis according to what/who a person wishes to get to know. A graphic depiction of this journey, designed by Terry Irwin from a description of the Goethean process by Margaret Colquhoun, is shown in Figure 5. Distinctions between different parts of the process are not rigidly specified but can vary according to emphasis and experience.*

* The outline of Goethe's methodology that I present here comes primarily from work with Margaret Colquhoun, who is a visiting teacher on the MSc at Schumacher College.

1. FIRST IMPRESSIONS

The initial step is an encounter that seeks to be free of all preconceptions, in the way a child engages with a new experience, being open and receptive. This is an intuitive perception of the phenomenon as a whole, without analysis or ideas, based on the initial encounter only. It is carried out in silence and by the individual. Clearly if the phenomenon is familiar, this innocence is difficult to achieve, but it is sought. This becomes part of the training.

4. Seeing in Beholding

Asking *who are you,*
listening to what it tells you
about itself

6. Developing the Idea

The idea arises out of our
merging with the essence
of the thing

3. Flowing in Time

Imagining how it *came to
be,* letting the past flow
into the present and imaginging the future

**5. Becoming One
with the Essence**

An intuitive synthesis
of the previous steps in
order to recognize the
essence

7. Growing the Plan

Developing the idea
further, using an integrated approch of intuition
and observation

**2. Exact Sensorial
Perception**

Objective observation
of *what is there*

8. Landing the Plan

Objective observation
of *what is there*

1. Intuitive Meeting

Developing a *sense
of feeling* for the thing
observered, without
judgement or analysis

Past ⟶ Now ⟵ Future
Science/Fact   *Art/Creativity*

Figure 5. Goethe's methodology depicted as a series of stages in the process of a holistic encounter with another being, joining science with art in a journey from past to future. (Drawn by Terry Irwin, Schumacher College.)

2. EXACT SENSE PERCEPTION

This involves detailed sensory examination of the phenomenon: look, taste, listen, touch, smell; describe the details in factual language. This empirical study can involve drawing, taking notes, verbal descriptions and other observations that are shared in a group to construct a collective account. No ideas or preconceptions are used or discussed in this sensory encounter with the phenomenon.

3. EXACT SENSORIAL FANTASY

The observational details gathered in the previous step are used to re-experience the whole and its parts through imaginative reconstruction, bringing it into being in the imagination. The qualities of the phenomenon are experienced at this stage, not by inventing them but by experiencing them intuitively. It is often helpful to draw, sculpt or otherwise represent the being and live into its process and form, allowing it to speak.

At this stage it is appropriate to reconstruct the history and the wider context of the being, looking at photographs of a tree or a landscape or a building, as we would look at photographs of a person and their family to get to know them better. If a landscape and/or buildings are under study, the geological history of the place and whatever historical records exist are useful in getting a full picture of historical context. It is useful to compare these details and one's own experience with those of others, in order to seek consensus and to recognize one's own idiosyncratic responses to the phenomena one seeks to understand more deeply. Recognition of personal biases is part of the training involved in seeing the other.

4. SEEING IN BEHOLDING

This is the step where the other being is encountered and recognized in its distinctive nature, both in itself and in its relationship to the broader context. There can be a deep feeling of the being in its place, the intrinsic value and meaning of another being or the *genius loci* of a woodland or a landscape. The place expresses itself through the insight of those participating in the work.

5. BECOMING ONE WITH THE ESSENCE

At this point the being is experienced in such a way that knowing is combined with an ethical relationship in a unified *scientia intuitiva* that leads towards responsible creative action. It is the moment of bonding in full recognition of the other. This experience is often not distinctly separated from the previous process, but tends to fulfil it.

6. GROWING THE PLAN

In this stage it is possible to consider any actions, appropriate to the experience, that may facilitate the expression of meaning elicited by the being. This can take any one of a number of forms of expression, such as drawing, sculpting, writing, designing a building appropriate to the place, or dancing the experience of the being.

7. ASSESSING THE PLAN

Experience and evaluate the quality of the new being that has co-arisen (its form, function, economy and elegance, beauty, health, appropriateness to context). Assess the value of the new whole that has come into being. Again this involves all members of the participating group, looking for consensus concerning right action.

These procedures of Goethean science can be combined at any point with those of Consensus Methodology described in Chapter 3, where the question of including 'subjective' evaluations in scientific method and giving them intersubjective ('objective') status was examined. The experience of the participating group in evaluating the phenomena can be examined for reliability using the statistical methods of Free Choice Profiling and Generalized Procrustes analysis, as developed by Wemelsfelder (2000) and her colleagues. Tom Butterworth, a previous MSc student, has described this procedure in a study of the qualities of three different woodlands on the Dartington Hall Estate in Devon, where Schumacher College is situated. In general, these methods can be applied in any context where a community seeks consensus for action appropriate to local context. In Chapter 7 I shall describe the comprehensive approach developed by Christopher Alexander (2003) to appropriate architectural construction in different landscapes, relating it to Goethean procedures. There is a remarkable convergence here of principles regarding the perception of form, the needs of nature and of culture, and issues of right action in responding to need. This extends to the whole of human design, which is explored also in the next chapter.

Holistic realism

Goethe was a realist in his scientific perspective: nature exists and expresses herself in forms that exist in embodied reality. There is no matter without form and no form without matter, but matter now includes experience or mind as an aspect of its being, which is always in the process of becoming. The fusion of subject and object in a moment of experience in Goethe's description of engaged attention anticipated the view developed in quantum mechanics wherein the observer cannot be separated from what is observed: the two have to be understood as a unity. When the philosopher A.N. Whitehead was asked the classic naïve realist question whether the rose really is red, or just seems red to us, his answer was that the two situations are entirely different. The colour and the beauty of the rose are real aspects of a social world with humans and roses; they are not simply located in the rose, any more than the colour red is simply located in white light. The whole situation has to be taken into account, and all the relevant permutations required to reach a true understanding of the phenomena, according to Goethe's scientific method.

Goethe certainly changed his views during his long life, and there is no way in which anyone can present a definitive, fixed interpretation of what he believed. Furthermore, this isn't the point, which is rather to build on what Goethe points towards as a perspective on nature that helps us to heal the fractured relationships into which we have fallen. Since the world is one of creative process, our job is to engage with the present in a way that allows us to take right action by recognizing the reality and the rights of all other beings, as Goethe insisted in his conversations with Schiller. Human freedom is not a sufficient foundation for right action in our present circumstances. However, the challenge of articulating a foundation for a fully embodied engagement with ourselves and nature is daunting, possible only because contemporary voices are speaking with a common visionary insight concerning the nature of the Great Work in which we need to participate.

7. Living the Great Work

In this concluding chapter I shall explore pathways of transformation in which we may participate to bring about a radical shift of focus and activity in our culture, from a position of dominance and control on the planet to one of appropriate participation. Many voices are engaged in this enterprise which has the form of a radical transition, like a phase transition from one state of matter to another that lies a hairsbreadth away but expresses a very different way of being in the world. This type of change has been characterized in the past as an engagement with the Great Work, the *Magnum Opus* of the alchemists who used the symbolism of base metals transmuting into gold in the practitioner's crucible. However, the alchemist's laboratory was a place of labour *and* a place of prayer, *labor* and *oratorium*, since no transmutation would occur in the crucible unless the worker simultaneously underwent transformation from a lower to a higher, more responsible form of participation in the process. It is this engagement with the changing worldview that is emerging in our culture, and the practical work of shifting the focus of our activities in simple, direct and realizable ways, that I wish to explore here.

Ways of knowing

September 11, 2001, has already entered human consciousness and the annals of our time as a turning point in global culture, the apogee of a particular development symbolized by the Twin Towers of the World

Trade Centre in New York. They expressed unlimited confidence in the power of democracy, human rights, economics and trade to transform all cultures according to the vision that had emerged in 'the West.' Their sudden, dramatic destruction in an extraordinary attack by an alienated cultural group that saw these as symbols of an evil power that had gripped the West and was threatening the world was the most dramatic wake-up call by humans to humans that the new twenty-first century is ever likely to experience. There will be plenty of more dramatic events in this century that follow indirectly from human activity, but no cultural actions are likely to equal this in its significance.

I have a dramatic memory of the news of this event entering my awareness. At Schumacher College in Devon, where I learn and teach, a group of us had been on a field trip that day to Dartmoor, a bit of local wilderness in the south-west. We returned late in the afternoon to the College to be told the news, and then to watch it, shocked but not entirely surprised, on TV. The following morning we gathered, a group of nine students and three faculty participating in the MSc in Holistic Science, for a philosophy class. The leader of the class was Jordi Pigem from Barcelona, who taught philosophy at the College. He lit a candle, read a poem by the Buddhist monk Thich Nhat Hanh entitled 'Call Me By My True Name,' and asked 'How did we get here?' — the question that was on everyone's mind. The searching discussion that followed led to an image that encapsulated the deeper aspects of the questions we had been pursuing in Jordi's philosophy classes, and took us beyond into an emergent realization. We were acutely aware that the way of knowing the world developed by Western science was a very limited, though powerful one, driven by the desire to gain control over nature. This works, but gains control through abstraction and reduction, which separates the knower from the known and tends to alienate rather than unite them through an empathic relationship of respect and acknowledgement. This is the upper loop in Figure 6 shown here. The process is driven by fear and suspicion of the unknown.

We realized that there is another way of gaining knowledge of the world that depends on participation. This is the Goethean way of direct knowing that recognizes the other as a legitimate and unique being, an 'I-Thou' relationship as Martin Buber would have described it. Instead of abstracting and homogenizing nature into general categories, this recognizes uniqueness, difference, and diversity as the expression of creativity. The result is a sympathetic union of the knower and the

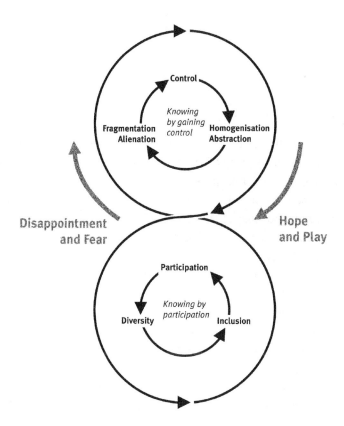

*Figure 6. Depiction of two ways of knowing: one through gaining control,
the other through participation. Both are connected, with transitions
driven by fear, or by hope and play mediated by love.*

known without losing their distinctness. This process of knowing
through participation is driven by love and trust of the real. The loop
of fear and the loop of love are connected dynamically, with transitions
from one to the other occurring when confidence in knowing by control
results in hope and play, which takes the person into the participation
through love loop; or conversely, disappointment and fear can shift a
person out of participation into control. Fear is a legitimate feeling
that needs to be respected and honoured, but mustn't dominate continuously. Love can be reached and expressed in participation, but it
doesn't last forever. The loops are reflective of the Yin-Yang dynamics

of ancient China or the Brahma-Kali polarity of the Hindu cosmology, which inevitably arise in cultures and have emerged in our own, with its own distinctive qualities.

Gaia reborn

What are the signs that new forms of awareness are emerging from science itself, with the possibility of taking us out of the control and into the participatory mode of relating to nature? Shifting a culture's perspective on the world is no easy task. It is widely known that when Galileo tried to convince his contemporaries to abandon scholastic knowledge from books and use the scientific method of direct observation and measurement of natural processes he met deep opposition from the Church because he was challenging received truth. He was forced to recant his belief that the Earth *actually,* rather than virtually in theory, moves round the Sun rather than the Earth being stationary at the centre of the planetary system.

There is an interesting contemporary story that has resonances with Galileo's experience in the seventeenth century. In the 1960s the scientist/inventor James Lovelock, who was working with NASA on issues relating to extraterrestrial life, had the insight that the composition of the Earth's atmosphere distinguishes it from other planets in a manner that tells us something profound about the relationship between living organisms and their inorganic environment. He wrote an article in the science journal *Nature* (1965) arguing that life doesn't simply adapt to given conditions on the planet where it takes root; it changes those conditions and stabilizes them so as to perpetuate itself, like a living organism. Lovelock was of course very excited about this idea and in 1969 he went for a walk with a neighbour of his in Cornwall, the writer and Nobel laureate William Golding, to whom he explained his theory. Golding recognized the remarkable significance of this concept of a living Earth and said that a big idea needed a big name: *Gaia.* Lovelock at first misunderstood what Golding was suggesting, then heard the name of the Greek divinity and recognized its power in relation to his insight into the qualities of life on Earth as an active agent in regulating the properties of the planet.

Lovelock's new theory, supported and extended by biological evidence from the work of Lynn Margulis on the power of microbes to alter planetary conditions, was presented to the scientific world as the Gaia Hypothesis in a 1974 article co-authored by Lovelock and Margulis in the journal *Tellus*. Here was science based on sound evidence but dressed in the garb of the ancient Greek goddess of the earth. Gaia was being reborn, emerging within Western science at about the same time as Chaos was being rediscovered as a creative principle. Here were two of the three principles of the ancient Orphic Trinity re-emerging in science. What did the scientific community make of this? They cast the Gaia hypothesis into the outer darkness. Why? Because Lovelock and Margulis had violated not one but two principles of orthodox science.

The first violation was the suggestion that there are basic aspects of evolution that do not conform to Darwinian principles: according to the Gaia hypothesis, life does not simply adapt to given geophysical conditions on earth but can change them so that they are more suitable for life on the planet. For instance, microbes can alter the composition of the atmosphere (CO_2, NH_3, CH_4, O_2) so that it is more hospitable for life than conditions on 'dead' planets, particularly by ensuring that hydrogen remains on the planet, largely in the form of H_2O in the oceans, the cradle of life. They can also stabilize the salt composition of the oceans, regulating calcium by deposition in microbial skeletons, and planetary temperature in ranges suitable for life by affecting cloud cover. The whole earth thus takes on the dynamic characteristics of a living organism, regulating its own vital variables as an organism does.

The second violation was the use of the term 'Gaia' in the hypothesis, which implied that the Earth itself is a kind of living being rather than a set of blind, mechanical processes of the type that science recognizes to carry out planetary activity. This image of Earth as goddess was an especially powerful one for environmental activists, who were protesting the pillage of Earth's natural resources and the pollution of land, sea and air by (for example) excessive burning of fossil fuels and the anger that many people were experiencing when they realized what we had done and are still doing to our planet. The New Age activists simultaneously embraced the third principle of the Orphic trinity, Eros, the force that through the green fuse creates the flower and the rest of the world of living form, as we have seen in Goethe's life and work.

Gaia's chief proponent, Lovelock, was effectively excommunicated from the church of science as a result of the heresies in the hypothesis. He responded with a vigorous defense of his first heresy, agreeing that Darwin's principle of natural selection was one of the mechanisms of evolution but insisting that life itself changed conditions on Earth as well as adapting to change. This principle has now been accepted by the scientific community, which uses the term 'Earth System Science' to describe the extended picture of terrestrial evolution that Lovelock and Margulis introduced. But acceptance came at a price: Lovelock recanted with respect to the implication that Earth has any qualities of intention or caring for life on the planet, a notion that was seen as a kind of animism, which is absolutely forbidden in science. It is all very well using animistic metaphors in science writing, as Lovelock has done in his latest book *(The Revenge of Gaia,* 2006), but these are not to be taken literally. The church of science had spoken.

We have seen that the dialectic of scientific understanding is now moving to the point where the taboo on animism is beginning to erode under the combined need to recognize the reality of our primary experience in feelings and the logical consistency required to account for their evolutionary origins from 'dead matter.' We will not have satisfactory solutions to these issues for some time, which will be the stories of a new age. The urgent issues revolve around finding ways forward without knowing what the outcomes will be. How are we to proceed with the task of healing ourselves and the Earth without being able to predict the result of our actions? This is precisely where new and old insights come together in an understanding that the reliable journey depends on feeling our way without knowing the outcome. This means putting into operation the principles of 'living on the edge of chaos,' which implies living in social networks with non-hierarchical connectedness: maximum freedom to the individual, maximum order to the collective. Humanity is now evolving from the hierarchical structure of industrial culture to the network structure of robust, creative, locally empowered societies of beings. We are learning how to act effectively, creatively, within such a global network through local action whereby people use their local knowledge and their intuition to effect appropriate transformation. This is the beginning of the Great Work, based on a new mythos of the living, creative cosmos impregnated with feeling and meaning as well as intelligible self-consistency. The new mythos is actually the re-emergence in contemporary culture of the Orphic Trinity of Chaos, Gaia, and Eros. The first two developed dialectically from within science in the 1960s, as we have seen. Eros also emerged in the Sixties in Hippie culture with its emphasis on the healing power of music and love. Orpheus, the legendary musician, physician and spiritual healer, was suddenly among us. It has taken several decades for this powerful experience to become grounded in practical lifestyles, but that is what is now emerging.

Living for now, not the future

There is an apparent paradox within this movement towards a new mythos and culture: the focus of the Great Work is on *local* action that realizes a collective, cooperative vision, but this action is based on *universal* principles of coherent behaviour that we learn from observation of nature. The paradox is resolved by recognizing that these principles apply to the *process* whereby coherence arises, not to some abstract, transcendental or ideal features of the *state* towards which we are moving. The point is not to describe the meaning of life or the universe as a whole, nor to describe some future ideal condition with which participants agree as a goal for present action. The objective is to discover lives of meaning that are necessarily embedded within a specific natural context, based on deep local knowledge of place and history. Right or skilful action is then expressed through local narrative and action, not through abstract principles. Since the cosmos is one, cohesion will emerge through local attention to detail and relationship in practical action. This is something that one discovers rather than imposes; it is an emergent property of proper relationships that cannot be known in advance, though it is recognized in retrospect. The trick is to live in the present, to pay attention to detail and maintain the integrity of the process in which you are engaged, informed by a love that recognizes conflict and is prepared to live through it to resolution. Then the future will look after itself and have the right qualities (cf. Shaw 2002, for a beautifully coherent description of how this can work in the corporate sector).

There is a sharp contrast between being present with others in current action, and the emphasis of our industrial culture on a future that is constantly offered as an improvement on the present but of course never arrives. We live for an unreal illusion, however perfect it may seem in a fictive virtual reality or an ideal society, and we are blinded to the destructive ugliness of our present reality. We have invented time as a scarce resource that we struggle for but never have enough of, with the promise that by mortgaging our present lives we will realize a wonderful future. The process of continuous growth that our politicians and economists offer as the path to happiness and fulfilment is in fact a policy of conflict resolution that continuously transfers our debts to nature, whose bounty we are living from and systematically destroying. We are now emerging from this dream, that is turning into a nightmare.

There is a tension that arises in the global market place. A plethora of brands and choice, all competing for the consumer's attention, create confusion, a continuous search for gratification and a kind of ennui in the shopper. This is an aspect of the disease of consumption in industrialized cultures known as 'affluenza.' The stresses of remaining individual through projection and fashion result in bewilderment about identity, which is what choice of goods as a form of individual projection of the person was initially supposed to allow and encourage.

Many voices are calling our attention to the contradictions we have fallen into. Marion Woodman expresses our problem in powerful archetypal terms:

> Matter is accumulating in heaps all around us, more and more material. We can't get enough. We are burying ourselves in it, whether in possessions or flesh. We rape nature, the Great Mother, with very little sense of guilt. The devouring mother is taking over but we don't open our eyes to see her. The anorexic girl unconsciously says a flat NO to the witch, but is unconsciously devoured. The fat girl, caught between her hatred of the witch outside and the witch within, builds her own fortress in an effort to escape. The alcoholic escapes through his trickster spirit. Meanwhile the Eros principle, the spirit of love, cannot relate to the daily ravaging of human instincts. The love that once existed between nature and man has almost been annihilated. It is at the point of death, indeed, that new life can appear. Yet the archetype of the feminine, as it is constellating now, is not clear. Maybe the darkness is not yet deep enough. (Woodman 1982)

Sam Keen observes:

> I suspect that the major cause of the mood of depression and despair and the appetite for violence in modern life is the result of masses of people who are enslaved by an economic order that rewards them for labouring at jobs that do not engage their passion for creativity and meaning. (Keen 1999)

Natural economics

We know that the economic system within which we live our lives has both its strengths and its weaknesses, and it is difficult to imagine trading processes that are radically different in their essence. However, that is precisely what we need to do, and there are imaginative proposals by contemporary thinkers to consider. The one that I shall now describe comes from the work and experience of Bernard Lietaer, whose books *The Future of Money* (2001) and *Access to Human Wealth: Money beyond Greed and Scarcity* (2003) present realistic economic visions based on a thorough acquaintance with the present dominant money system. Lietaer was Professor of International Finance at the University of Louvain in his native Belgium, where he was also head of the Organization and Planning Department at the Belgian Central Bank. After extensive experience in the world of finance, he has now become a Research Fellow at the Centre for Sustainable Futures in Berkeley, California. He sees money as a cultural artefact that is ours to design appropriately, in accordance with whatever vision we may have of nature and culture, creativity and trade. While economists claim that corporations are competing for markets and resources, Lietaer sees them competing for money, using markets and resources. In his words:

> I believe that greed and competition are not a result of immutable human temperament; I have come to the conclusion that greed and fear of scarcity are in fact being continuously created and amplified as a direct result of the kind of money we are using.
>
> For example, we can produce more than enough food to feed everybody, and there is definitely enough work for everybody in the world, but there is clearly not enough money to pay for it all. The scarcity is in our national currencies. In fact, the job of central banks is to create and maintain that currency scarcity. The direct consequence is that we have to fight with each other in order to survive.*

---

* This extract and the following quotes come from an interview by Sarah van Gelder with Lietaer that is available at: http://www.transaction.net/press/interviews/lietaer0497.html

Here is the Darwinian worldview of survival of the fittest through competition, expressed in our economic system. Darwin saw nineteenth century market capitalism as the path of cultural progress, and believed that it must also express the essential principles of progressive evolution. In the twentieth century Darwinian natural selection based on competition was seen to be an essential truth of nature that should be embodied in cultural evolution, so we get the principles of economic development presented as natural truths. This illusion results from a play of mirrors. There is no need to take either the economic or the evolutionary model as the essential truth, and good reason to look for more complete ways of seeing nature and culture. Lietaer looks deeply into cultural history and the human condition to find an alternative model for economics, as follows:

> My analysis of this question is based on the work of Carl Gustav Jung because he is the only one with a theoretical framework for collective psychology, and money is fundamentally a phenomenon of collective psychology.
>
> A key concept Jung uses is the archetype, which can be described as an emotional field that mobilizes people, individually or collectively, in a particular direction. Jung showed that whenever a particular archetype is repressed, two types of shadows emerge, which are polarities of each other.
>
> For example, if my higher self — corresponding to the archetype of the King or the Queen — is repressed, I will behave either as a Tyrant or as a Weakling. These two shadows are connected to each other by fear. A Tyrant is tyrannical because he's afraid of appearing weak; a Weakling is afraid of being tyrannical. Only someone with no fear of either one of these shadows can embody the archetype of the King.
>
> Now let's apply this framework to a well-documented phenomenon — the repression of the Great Mother archetype. The Great Mother archetype was very important in the Western world from the dawn of prehistory throughout the pre-Indo-European time periods, as it still is in many traditional cultures today. But this archetype has been violently repressed in the West for at least five thousand years starting with the Indo-European invasions — reinforced by

the anti-Goddess view of Judeo-Christianity, culminating with three centuries of witch hunts — all the way to the Victorian era.

If there is a repression of an archetype on this scale and for this length of time, the shadows manifest in a powerful way in society. After five thousand years, people will consider the corresponding shadow behaviors as 'normal.'

The question I have been asking is very simple: What are the shadows of the Great Mother archetype? I'm proposing that these shadows are greed and fear of scarcity. So it should come as no surprise that in Victorian times — at the apex of the repression of the Great Mother — a Scottish schoolmaster named Adam Smith noticed a lot of greed and scarcity around him and assumed that was how all 'civilized' societies worked. Smith, as you know, created modern economics, which can be defined as a way of allocating scarce resources through the mechanism of individual, personal greed.

If greed and scarcity are the shadows, what does the Great Mother archetype herself represent in terms of economics? Lietaer answers this in the following perceptive way:

Let's first distinguish between the Goddess, who represented all aspects of the Divine, and the Great Mother, who specifically symbolizes planet Earth — fertility, nature, the flow of abundance in all aspects of life. Someone who has assimilated the Great Mother archetype trusts in the abundance of the universe. It's when you lack trust that you want a big bank account. The first guy who accumulated a lot of stuff as protection against future uncertainty automatically had to start defending his pile against everybody else's envy and needs. If a society is afraid of scarcity, it will actually create an environment in which it manifests well-grounded reasons to live in fear of scarcity. It is a self-fulfilling prophecy!

Also, we have been living for a long time under the belief that we need to create scarcity to create value. Although that is valid in some material domains, we extrapolate it to other domains where it may not be valid. For example,

there's nothing to prevent us from freely distributing information. The marginal cost of information today is practi-
cally nil. Nevertheless, we invent copyrights and patents in
an attempt to keep it scarce.

The question then arises how to achieve a different form of economics
which reflects the original meaning of *oekos,* the economy of the home.
Lietaer points out that the origin of the word 'community' comes from the
Latin *munus,* which means 'gift,' and *cum,* which means 'together, among
each other.' So community literally means to give among each other.
Community is therefore an extension of home and family, as a group of
people who welcome and honor each others' gifts, both given and received.
The natural expression of these relationships is through the use of local
currencies, which Lietaer predicts will be a major tool for social design
in the twenty-first century, if for no other reasons than employment. His
view is not that these local currencies will or should replace national currencies; rather, they are called 'complementary' currencies. The national,
competition-generating currencies will still have a role in the competitive
global market, but complementary local currencies are a lot better suited
to developing cooperative, local economies. Local currency creates work,
but there is a distinction between work and jobs. A job is what you do for
a living; work is what you do because you like to do it. Most jobs will
increasingly become obsolete, according to Lietaer, but there is still an
almost infinite amount of fascinating work to be done.

Lietaer defines a complementary currency as one that operates in
parallel with the conventional currency, like the dollar or the euro or the
yen, and fulfills some needs that the conventional currency hasn't satisfied. Many of these arise to solve problems such as care of the elderly,
local unemployment, the restoration of a community or getting young-
sters off drugs. They work equally well in Mexico City and in fishing
villages in Canada. They can be based on low-tech paper-based systems
in Berkeley, California, or on high-tech smart card applications in Asia.
They were designed for small groups of fifty people in Australia, a city
of 2.3 million people in Brazil or prefectures of ten million in Japan.

In the US, probably the best known local currency is the Ithaca
'Hours.' These are accepted at the farmers' market; and farmers can
use the local currency to hire someone to help with the harvest or to do
some repairs. Some landlords accept Hours for rent, particularly if they
don't have a mortgage that must be paid in scarce dollars. With a local

currency, it quickly becomes clear what's local and what's not. K-Mart will accept dollars only; their suppliers are in Hong Kong or Singapore or Kansas City. But Ithaca's local supermarket accepts Hours as well as dollars. Use of local currencies creates a bias toward local sustainability. We can look forward to the increasing emergence of gift economies with local currencies in which the concept of property is radically altered to collective ownership and sharing. This is of course a restoration of the commons that have been destroyed by the felt need to allay fear through individual ownership functioning as security. When the fear dissipates through trust in local communities the need for individual ownership and the whole movement towards intellectual property will transform into shared collective use. Local currencies will facilitate this fundamental shift in the way communities use money and trade with one another.

Natural design

Natural economics, following nature's principles of sustainable abundance without creating waste and damaging the environment, requires that we adopt principles of natural design in all our production processes. This is part of the new mimesis. Whatever we produce must be made from materials that are all recyclable, using a process that is energy efficient, as occurs in nature. But in addition we are called upon to create beautiful and functional forms with a diversity that expresses the uniqueness of individual creation and context, as occurs in the natural world This is the challenge of biomimicry, of learning from nature how to live sustainably (Benyus 1998). We are required not simply to satisfy our needs intelligently, which is what we have been doing throughout modernity, though within a limited understanding of intelligence that is dominated by a functional perspective on evolution. In addition, we must now create in a manner that respects natural principles of balance in the use of energy and materials while simultaneously enhancing the beauty of the world, as do other life-forms. We may be rediscovering a perennial truth: that the job of humanity on earth is to contribute to and to celebrate the creativity that is inherent in all things. This is the activity that emerges from a communion of subjects that is now taking root in diverse centres where natural design is being taught and practised.

Victor Papanek is a contemporary voice who sees design as something that humans engage in all the time, 'for design is the primary underlying

matrix of life' (Papanek 1994). The influence of our design activities
on our visual and material environment is deep and usually pernicious.
Terry Irwin, who has spent her life in the field of design, sees a deep
need for transformation in the whole approach to design:

> Everywhere we look in the 'built' world we see the work of
> designers. Much of it is toxic, ugly, dysfunctional, harmful
> to the natural environment and ultimately unsustainable. If
> design is responsible for so many problems in the world, then
> surely it has the potential to be the basis of their solutions.
>
> 'Bad' design is the physical manifestation of a Cartesian/
> mechanistic worldview whose roots can be traced to both
> the Scientific Revolution of the seventeenth century and the
> Industrial Revolution which made the mass-production of
> goods and services possible. Design process has mirrored
> scientific process in many ways and has become linked to
> the consumer marketplace which is based upon short-term
> financial results.
>
> We must rethink the role of design in the twenty-first cen-
> tury; to harness the power of design to solve big problems
> with large consequences.
>
> We need a more appropriate definition of design: one that
> communicates its meaning and potential and which encour-
> ages everyone to take responsibility for the forms brought
> forth by human beings.
>
> Designers need to better understand how the world works.
> To do this they need to look outside the field of design for
> inspiration, understanding, know-how and know-why.
>
> Nature has been designing beautifully, usefully and sus-
> tainably for 4.5 billion years. Discoveries in the 'new sci-
> ence' have revealed basic principles about how life works
> that have great relevance for design.
>
> Re-educating ourselves and designing a new curricu-
> lum for tomorrow's designers can be a powerful catalyst
> in achieving a sustainable world. Alternative educational
> structures and formats must be developed to allow young
> designers to learn in a different way and established/mature
> designers to re-think their role in a sustainable society.
> (Irwin 2005)

Form and architecture

Closely connected with the movement toward natural design is the impulse to design and construct sustainable buildings. David Orr is the author of one of the most powerful books to describe how our relationship to the earth has been transformed by our knowledge and its misuse. In the introduction to *Earth in Mind: On Education and the Human Prospect* (1994), he observes:

— Male sperm counts worldwide have fallen by 50% since 1938, and no one knows exactly why;
— Human breast milk contains more toxins than are permissible in milk sold by dairies;
— At death, human bodies contain enough toxins and heavy metals to be classified as hazardous waste;
— Similarly toxic are the bodies of whales and dolphins washed up on the banks of the St Lawrence River and the Atlantic shore;
— There has been a marked decline in fungi worldwide, and no one knows why;
— There has been a similar decline in populations of amphibians worldwide, even where the pH of rainfall is normal;
— Roughly 80% of European forests have been damaged by acid rain;
— US industry releases some 11.4 billion tons of hazardous wastes to the environment each year;
— Ultraviolet radiation reaching the ground in Toronto is now increasing at 5% per year.

Those observations were made more than ten years ago and a lot more could now be added to the list. However, some things get better, such as the increase in the ozone levels in the upper atmosphere, a result of the decreased use of chemicals such as CFCs (chlorofluorocarbons), the refrigerator gases that were identified as ozone destroyers and legislated against in the Montreal Accord. Humans can collectively and individually take appropriate action. David Orr is one of those who shows how this can be achieved in architecture as an area of human activity that can follow the principles of ecological design that he has done so much to promote. 'Aside from our use of language, the act of building is perhaps the most distinctive thing about humans. Building for us, clever apes, is not just a nesting instinct but the way by which we make our ideas and values manifest. The act of building is a form of language that puts us on public display.' Orr has put himself and his beliefs on public display with a new building on the Oberlin campus in the States, described in the *New York Times* as 'perhaps the most remarkable of a new generation of College buildings.'

In a later book (2004) that describes in detail the principles, practices and the travail that accompanies ecological design in building, Orr says:

> ... most people most of the time have strong feelings about beauty, order, harmony and at some level are wounded by their absence. The sense of beauty is not, in other words, simply in the eye of the beholder, it comes with our hard wiring. ... The expression of beauty changes in different places and times and the possibilities for creating it are not equal, but our built-in sense of beauty and place is expressed in many different ways. That sense of place, however, breaks down as the scale increases. And driven by population growth, industrialization and mechanization, the scale of human civilization has increased with astonishing speed in the past two centuries. Villages become cities, cities become metropolitan and then formless, sprawling, megapolitan regions. The word 'sprawl' doesn't quite describe what is more like an eruption of humanity fueled by easy access to ancient sunlight and the draw down of ecological capital of soils, forests, and biological diversity, which is to say, ecological disease. Sprawl brings, too, a range of human health problems rare in more concentrated societies. Suburbanites living in isolation from each other and dependent on the car for transportation, are more obese, suffer more often from heart disease, are more prone to asthma, and victims of other diseases rare in more concentrated communities. (Orr 2004)

Here we have again the recognition that individuals in their communities can reach consensus about creating forms appropriate to place and time so long as scale and communication are preserved.

Building for wholeness and beauty

Christopher Alexander is an architect whose direct experience of the objective beauty of natural form inspires his conception of the building process. In his immensely influential 1977 book, *A Pattern Language,* he describes the principles of natural, living form that inform good architectural design. In his latest work, a comprehensive and inspiring four-volume development of the ideas expounded in *A Pattern Language* entitled *The Nature of Order* (2002), Alexander presents fundamental principles of form. These are based on the primary concept of wholeness and how this is realized through living centres that organize and animate the whole, and illustrated through innumerable examples from art, architecture, science and nature. 'At each place in the world — with its natural habitat, ecology, buildings, materials, actions and events — there is, at any instant, some given wholeness, that is, some definite, well-defined system of centres that creates the organization of that part of the world.' *(Nature of Order* vol.1, p.106) Alexander demonstrates how good design reflects natural principles of form, whether animate or inanimate. Nature and culture both express themselves through forms of coherence and wholeness that run through all creative action, whether in natural process, in art or in science, which are all now reconnected as I described in Chapter 5, 'The Life of Form.' These principles transcend personal, subjective opinion and are embedded in physical and biological reality in forms that are accessible to human perception. As David Orr expresses it, beauty is not simply in the eye of the beholder. In *The Nature of Order,* Alexander presents his case for the presence of beauty as a real quality of the natural world and the possibility of its objective presence in our creations. Here is what he says in the volume entitled *The Luminous Ground:*

> What I call 'the I' is that interior element in a work of art, or in a work of nature, which makes one feel related to it. It may occur in a leaf, or in a picture, in a house, in a wave, even in a grain of sand, or in an ornament. It is not ego. It is not me. It is not individual at all, having to do with me, or you. It is humble, and enormous: that thing in common which each one of us has in us. It is the spirit which animates each living center. ... In some form, it is the personal nature of existence revealed in the building, that I am searching for. It is 'I,' the

I-myself, lying within all things. It is that shining something
which draws me on, which I feel in the bones of the world,
which comes out of the earth and makes our existence luminous. (Alexander 2002, vol.4, p.2)

Compare this with Wordsworth in 'Lines Composed a Few Miles
Above Tintern Abbey':

> ... And I have felt
> A presence that disturbs we with the joy
> Of elevated thoughts; a sense sublime
> Of something far more deeply interfused,
> Whose dwelling is the light of setting suns,
> And the round ocean and the living air,
> And the blue sky, and in the mind of man ...

Alexander belongs to the same tradition of Romantic realism that I
described in Chapter 6 through Goethe's life and work, to which of course
Wordsworth himself belongs. This movement has now reached the condition of integration with science through the union of quantities with qualities that characterizes phenomenology and holistic observation. Alexander
is articulating the impulse towards wholeness, coherence and skilful or
right action, inspiring new generations of builders and designers to understand that their job is to use natural principles to inform every step of their
creative activity. The connection between Goethe and Alexander is deep,
though it was not Goethe who inspired him. It was nature, and in particular the British countryside, that has been tended with care by farming
and country communities with a natural, pragmatic poetry in their veins
that comes from a deep understanding of natural process and how to live
in harmony with it. This is becoming undermined by industrial farming,
by biotechnology, and by developers whose sole inspiration is functional,
generating maximum agricultural and monetary output as required by the
dominant economics of competition and scarcity.

At Schumacher College we had a remarkable opportunity to learn how
Christopher Alexander searches for 'that shining something that draws him
on' in his architectural work when he led us in the process of designing
possible new structures appropriate to the land around the College and in
relation to existing buildings. He explores appropriate architectural design
with the community that will make use of the buildings through a process

that involves close attention to the quantities and the qualities of the site on which the buildings are to be constructed. An important aspect of this is the construction of a topographic, scale model of the land, with its existing contours, buildings and vegetation, which define the context for any new construction. We had spent the week before Alexander came to direct this process studying the qualities of the land surrounding Schumacher College with Goethean scientist Margaret Colquhoun, who has a profound understanding of this process through years of experience. This had brought us into a relationship of sensitivity to place that is essential to Alexander's way of approaching the site, effectively asking the land what would be appropriate to its nature. As a result, we were prepared in some measure for the engagement in the search for appropriate form in building design that is the heart of Alexander's genius. What we experienced was a design process that offered the possibility of a deeply authentic and participatory involvement of those engaged in it, giving us some insight into Alexander's principles of design based on the properties of living form with its wholeness, coherence, and capacity to heal. In some limited measure we travelled with him toward a goal that Alexander describes as the Ultimate I. 'To reconcile the vision of matter with the experience of personal relatedness we feel when in touch with living structure, I have, in addition, introduced the conception that living centres open some kind of window or tunnel to a ground *and to an ultimate I which constitutes the ground.*' Alexander challenges us with his vision of architectural principles that reconnect us to nature and to the beauty that resides in our souls. This is a significant part of the Great Work that sometimes seems like an unattainable goal but is in fact a necessity that we can now achieve.

Reformulating rights and responsibilities

The Great Work before us, the task of moving modern industrial civilization from its present devastating influence on the Earth to a more benign mode of presence, is not the role that we have chosen. It is a role given to us, beyond any consideration with ourselves. We did not choose. We were chosen by some power beyond ourselves for this historical task. We do not choose the moment of our birth, who our parents will be, our particular culture or the historical moment when we will be born. We do not choose the status of spiritual insight or political or economic conditions that will be the context of our lives. We are, as it were, thrown into existence with a challenge and a role that is beyond any personal choice. The nobility of our lives, however, depends upon the manner in which we come to understand and fulfil our assigned role.

Yet we must believe that those powers that assign our role must in that same act bestow upon us the ability to fulfil this role. We must believe that we are cared for and guided by these same powers that bring us into being.

Our own special role, which we will hand on to our children, is that of managing the arduous transition from the terminal Cenozoic to the emerging Ecozoic Era , the period when humans will be present to the planet as participating members of the comprehensive Earth community. This is our Great Work and the work of our children ...

These thoughts and visions of Thomas Berry in his testament *The Great Work: Our Way into the Future* (1999) define the job with which we are faced, and express the faith that we are up to it. One of the primary tasks we need to address is the question of the rights of the beings, other than humans, who occupy this planet with us and are entitled to live their lives without being damaged or destroyed. Berry points out that we tend to regard the American Constitution as an enlightened document that expresses advanced, liberating concepts of human rights that emerged during the revolutionary years in Europe, particularly the French Revolution with its abolition of monarchic and aristocratic privilege from the political process. The United States of America then became the embodiment of these principles of human freedom. However, these applied exclusively to humans, who were guaranteed participatory governance, individual freedoms and the right to own and dispose of property with no legal protection for the natural world. We rightly regard genocide as a crime against humanity, but exterminating other life forms on earth is legal. If we are to survive on this planet it is necessary to rethink these principles so that all the beings that constitute the earth are recognized as beings with rights, just as we have. How is this to be done?

Earth jurisprudence

In a book entitled *Wild Law: A Manifesto for Earth Justice* (2002), the environmental lawyer Cormac Culinan explores an extended notion of rights in law to cover other species as well as crucial components of the Earth as an integrated dynamic whole such as the atmosphere, mountains, rivers, lakes, oceans and soil. He refers to this as Earth Jurisprudence. Articulating and establishing these rights is one part of the journey that Thomas Berry encourages us to undertake, as he urges in the foreword to Culinan's book.

The primary principle stated by Culinan is based on a holistic view of the Earth.

> Within the Earth System the wellbeing of the planet as a whole is paramount. None of the components of the Earth's biosphere can survive except within the Earth ecosystem. This means that the wellbeing of each member of the Earth Community is derived from, and cannot take precedence over, the wellbeing of Earth as a whole. Accordingly the first principle of Earth Jurisprudence must be to give precedence to the survival, health and prospering of the whole Community over the interests of any individual or human society.

People are responding to this in a variety of ways, and networks of committed groups with objectives of different kinds, appropriate to place, are forming alliances to work within the overall vision.

Earth democracy

One of the most powerful contemporary voices urging the adoption of principles of equity for all planetary citizens, human and non-human, is that of the Indian ecological and social activist, Vandana Shiva. Her message is simple and direct: human society prospers best when it functions according to the principles that operate in the natural world, and it finds itself in increasing difficulty when these are replaced by human concepts that are based on concepts of scarcity, ownership and control. Trained in mathematics and physics, but then shifting her attention to ecology, biology and social issues, Shiva has developed a powerful critique of what is missing from conventional scientific ways of knowing, particularly in the life sciences. It became evident to her that the living realm operates on principles of creativity and coherence that involve life and death, creativity and dissolution, in ways that are totally absent from the way we think about our technological constructs and artefacts. Furthermore, it emerged that these principles of creativity and coherence belong not simply to living beings but to nature in general, as described in the perennial philosophies of East and West, particularly Buddhism. Shiva has articulated in a series of books (1988, 1993, 2003, 2005) the shift of worldview that is now taking place from the mechanical and reductionist to the organic and holistic, and the practical courses of action required to implement these.

This revolution in concept and praxis is taking place within a cultural context and in the face of institutions that tend to deny or contradict the new understanding, so we face conflicts at every turn. Shiva has been one of the most effective contemporary figures to make these issues clear and to engage with them. One of her most dramatic conflicts is with our economic system, which works with the concepts of growth and development as the processes that will bring progress and wealth to 'undeveloped' countries in the Third World. 'Growth' and 'development' both come from biology, but their meaning has been reversed in their new context, and in the process corrupted. In the organic world growth is inextricably linked to decay and recycling: that is, life is necessarily linked to death, defining the life cycles in which all organisms participate. Growth and decay are always in balance in natural systems, as otherwise there would be depletion of the resources for life. Development is an aspect of the life cycle: it is the emergence of the mature, fully-formed organism, the willow tree or the newt, from the seed or the fertilized egg. This is an intransitive process that proceeds from within and expresses the nature of the being that is undergoing development in a manner appropriate to its context, like the unfolding of an oak tree from an acorn.

In the economic system the opposite is the case: growth measured as GDP or GNP occurs only if there is something produced in addition to the process of making an object and recycling its constituents. A tree that grows and dies, that recycles, is not regarded as economic growth. There has to be also some value or wealth, measured as money or equivalent that can grow indefinitely, without bound. Furthermore, development becomes a transitive process that is directed and controlled from outside, not the emergence of a form that expresses the nature of the being, a culture or a country, as a self-organizing, internally-directed process. Economic development as defined by the IMF, the World Bank and the WTO is the process that 'undeveloped' countries should go through to reach the mature, industrialized state. The paradox here is that this concept of development results in a process that can continue only by destroying the natural world, the source of the 'externalities' that the continuous growth process depends upon. The fact that we have painted ourselves into a critical corner on the planet is now obvious to anyone wishing to look. Shiva looks and takes action, primarily in India but also in cooperation with activist groups throughout the world. Her latest book, *Earth Democracy* (2005), describes ways in which people are implementing the new/old vision of living sustainably and harmoniously through networks of local people engaged in production, preparation and distribution of food and other goods in ways that protect the health of the environment and of people as a whole, hence of the planet as a community. The key is going local and using sustainable practices that make use of all the technology currently at our disposal to implement the post-Industrial culture. Various centres and communities around the world are now engaged in this work, linked together as networks of local activists with a potentially global impact.

The great transformation

The Great Work, the *Magnum Opus* in which we are now inexorably engaged, is a cultural transformation that will either carry us into a new age on earth or will result in our disappearance from the planet. The choice is in our hands. I am optimistic that we can go through the transition as an expression of the continually creative emergence of organic form that is the essence of the living process in which we participate. Like the caterpillar that wraps itself up in its silken swaddling bands prior to its metamorphosis into a butterfly, we have wrapped ourselves in a tangled skein from which we can emerge only by going through a similarly dramatic transformation. In the world of insects, this transformation occurs as a result of a self-digestion, a meltdown of the caterpillar in which only a few crucial foci of living tissue, the imaginal discs, remain intact. It is from these that the legs, wings, antennae, body segments and other structures of the adult form emerge as an integrated, transformed being, the butterfly. What the cultural correspondences of this metaphor might be we can only speculate. I see the imaginal discs, the living centres of the new age, already forming as those communities engaged in a transition to renewable energy resources, new technologies, and new patterns of cooperative engagement between humans and with the natural world in which there is no longer a distinction between nature and culture. These communities will have the distinctive qualities of their local contexts in which people experience rich lives of meaning and the abundance that arises from well-tended habitats based on ecological principles of diversity and mutualistic cooperation. This Gaian Renaissance will lead to what Thomas Berry calls the Ecozoic Age, in which all inhabitants of the planet are governed by principles of Earth Jurisprudence in an Earth Democracy. We can all experience lives of meaning and know that indeed there is no truth beyond the magic of creative participation in the life of the cosmos that embodies the liberation of Chaos, the abundance of Gaia, and the love of Eros.

References

Abraham, R. (1994) *Chaos, Gaia, Eros.* Harper: San Francisco

Alexander, C. (1977) *A Pattern Language. Towns, Buildings, Constructions.* Oxford University Press, NY

——, (2002) *The Nature of Order: An Essay on the Art of Building and the Nature of the Universe.* Vols. 1, 2, 3 and 4. Center for Environmental Structure: Berkeley, CA

Bak, P. (1996) *How Nature Works.* Springer: New York

Barabasi, A-L. (2003) *Linked.* Penguin Books: New York

De Beer, Sir Gavin (1971) *Homology: An Unsolved Problem.* Oxford University Press, UK

Ben Jacob, E., Shapira. Y., and Tauber, A. I. (2006) 'Seeking the foundations of cognition in bacteria; from Schroedinger's negative entropy to latent information.' *Physica A* 359, pp.493–524

Berkes, F., Colding, J., and Folke, C. (2003) *Navigating Social-Ecological Systems: Building Resilience for Complexity and Change.* Cambridge University Press, UK

Black, D.L. (1998) 'Splicing in the inner ear: a familiar tune, but what are the instruments?' *Neuron* 20: pp.165–68

Bockemuhl, J. (1998) 'Transformations in the foliage leaves of higher plants,' pp.115–28 in *Goethe's Way of Science; A Phenomenology of Nature.* D. Seamon and A. Zajonc (eds). State University of New York Press

Bortoft, H. (1996) *The Wholeness of Nature: Goethe's Way toward a Science of Conscious Participation in Nature.* Lindisfarne Press, US; Floris Books, Edinburgh

Branthwaite A., and Cooper P. (1981) 'Analgesic effects of branding in treatment of headaches.' *Br Med. J Clin Res Ed.* 282: pp.1576–78

Brody, H. (2001) *The Other Side of Eden.* Faber and Faber: London

——, (2002) 'The language of community.' *The Ecologist* 32, pp.36–38

Brumfiel, G. (2004) 'What's in a name?' *Nature* 430, pp.498–99

Buchanan, Mark (2000) *Ubiquity.* Weidenfeld and Nicolson: London

Burkhart, G.M. (2005) *The Genesis of Animal Play: Testing The Limits.* MIT Press

Camazine, S., Deneubourg, J-L., Franks, N.R., Sneyd, J., Theraulaz,
    G., Bonabeau, E. (2001) *Self-Organization in Biological Systems.*
    Princeton University Press

Caws, P. (1988) *Structuralism: The Art of the Intelligible.* Humanities Press
    International

Chalmers, D.J. (1996) *The Conscious Mind: In Search of a Fundamental
    Theory.* Oxford University Press: New York

Chomsky, N. (1968) *Language and Mind.* Harcourt, Brace and World: New York

Cilliers, P. (1998) *Complexity and Postmodernism.* Routledge: London

Clark, A. (2000) 'Phenomenal immediacy and the doors of sensation.'
    J. Consciousness Studies 7, pp.21–24

Clarke, C. (2005) 'Being and Field Theory.' *J. Consciousness Studies* 12,
    no.4–5, pp.135–139

Cole, B.J. (1991) 'Is animal behaviour chaotic? Evidence from the activity of
    ants.' *Proc. R. Soc.* London B 244, pp.253–259

Colquhoun, M. and Ewald, A. (1996) *New Eyes For Plants.* Hawthorn Press, UK

Collier, A. (1994) *Critical Realism.* Verso: London

Costa, M., A.L. Goldberger and C.-K. Peng (2002) 'Multiscale entropy analy-
    sis of complex physiologic time series.' *Phys. Rev. Lett.* 89, pp.68–102

Damasio, A. (2003) *Looking For Spinoza: Joy, Sorrow and the Feeling Brain.*
    Harcourt, Inc.: Orlando

Dawkins, R. (1986) *The Blind Watchmaker.* Longman Scientific and Technical:
    Harlow, UK

Delisi, C. (1988) 'The Human Genome Project.' *Amer. Sci.* 76, pp.488–93

Douady, S. and Couder, Y. (1996) 'Phyllotaxis as a Dynamical Self-
    Organizing Process.' *J. Theoret. Biol.* 178, pp.255–312

Ferrer y Cancho, R., and Solé, R. (2003) 'Least effort and the origins of scal-
    ing in human language.' *Proc. Nat. Acad. Sci. US* 100, 788–791

Franks, N.R., Bryant, S., Driffith, R. and Hemerik, L. (1990) 'Synchronization
    of behaviour within the nests of the ant Leptothorax acervorum.' *Bull.
    Math Biol.* 52, pp.597–612

Gehring, W.J., and Ikeo, K. (1999) '*Pax6.* Mastering eye morphogenesis and
    eye evolution.' *Trends Genet.* 15371–377

Gibson, J.J. (1979) *The Ecological Approach to Visual Perception.* Houghton
    Mifflin: Boston

Gleick, J. (1987) *Making a New Science.* Viking: New York

Globus, G.G., Pribram, K. H., and Vitiello, G. (eds) (2004) *Brain and Being:
    At the Boundary Between Science, Philosophy, Language and the Arts.*
    John Benjamin Publishing Co.: Amsterdam

Goldstein, J. (1999) 'Emergence as a construct.' *Emergence* 1, pp.47–62

Goodwin, B.C. (1972) 'Biology and Meaning,' pp.259–75 in *Towards a Theoretical Biology* (ed. C.H. Waddington) Vol.4, Edinburgh University Press

——, (1994) *How the Leopard Changed Its Spots.* Weidenfeld and Nicolson: London

——, (1999) 'Reclaiming a life of quality.' *J. Consciousness Studies* 6, pp.229–35

——, (2006) 'Understanding the origins, stability and frequency of biological forms.' *Harvard Companion to Evolution,* eds. M. Ruse and J. Travers. Harvard University Press

Griffin, David Ray (1998) *Unsnarling the World Knot: Consciousness, Freedom, and the Mind-Body Problem.* University of California Press: Berkeley

Griffin, Douglas (2002) *The Emergence of Leadership: Linking Self-Organization and Ethics.* Routledge: London and New York

Guelzim, N. Bottani, S., Bourgine. P., and Kepes, F. (2002) 'Topological and causal structure of the yeast transcriptional regulatory network.' *Nature Genetics* 31, pp.60–63

Hartshorne, C. (1972) *Whitehead's Philosophy: Selected Essays,* 1935–1970. Lincoln: University of Nebraska Press

Halder, G., Callaerts, P., and Gehring, W.J. (1995) 'Induction of ectopic eyes by targeted expression of the *eyeless* gene in Drosophila.' *Science* 267, pp.1788–92

Heidegger, M. (1958) *The Question of Being.* College and University Press: New Haven CT

Ho, M-W. (1993, 1998) *The Rainbow and the Worm.* World Scientific Publishing Co

——, (1998) *Genetic Engineering: Dream or Nightmare?* Gateway: Dublin

——, (2003) *Living With the Fluid Genome.* Institute of Science in Society: London

Husserl, E. (1960) *Cartesian Meditations: An Introduction to Phenomenology.* Allen and Unwin: London

Irwin, T. (2005) Excerpt from PhD research, University of Dundee

Ivanov, P.C., Rosenblum, M.G., Peng, C-K., Mietus, J., Havlin, S., Stanley, H.E., and Goldberger, A.L. (1996) 'Scaling behaviour of heartbeat intervals obtained by wavelet-based time-series analysis.' *Nature* 383, pp.323–27

Jensen, Derrick (2000) *A Language Older Than Words.* Context Books: New York.

Jeong, H., Mason, S.P., Barabasi, A.-L., and Oltvai, Z.N. (2001) 'Lethality and centrality in protein networks.' *Nature* 411, p.41

Kauffman, S.A. (1993) *Origins of Order: Self-Organization and Selection in Evolution*. Oxford University Press: New York

——, (1995) *At Home in the Universe*. Oxford University Press: New York

——, (2000) *Investigations*. Oxford University Press: New York

Keen, S. (1999*) Learning to Fly*. Broadway Books: New York

Keller, E.F. (1983) *A Feeling for the Organism*. Freeman: New York

——, (2000) *The Century of the Gene*. Harvard University Press

Kuhn, T. (1970) *The Structure of Scientific Revolutions*. University of Chicago Press

Langer, S. (1988) *Mind: An Essay on Human Feeling*. John Hopkins University Press

Leake, J. (2006) 'Pesticide Nun.' *The Ecologist* 36, pp.50–57,

Lietaer, B. (2001) *The Future of Money*. Random House

——, (2003) *Access to Human Wealth: Money beyond Greed and Scarcity*. Access Books

Lindberg, D. (1976) *Theories of Vision from Al-Kindi to Kepler*. University of Chicago Press

Lock, M. (1993) *Encounters with Aging: Mythologies of Menopause in Japan and North America*. University of California Press: Berkeley

——, (1998) 'Menopause: lessons from anthropology.' *Psychosom Med.* 60: pp.410–19

Lorentz, E.N. (1963) 'Deterministic non-periodic flow.' *J. Atmos. Sci.* 20: pp.130–141

——, (1991) 'Dimensions of weather and climate attractors.' *Nature* 353: pp.241–44

Lovelock, J. E. (1965) 'A physical basis for life detection experiments.' *Nature* 207, p.568

Lovelock, J. E. (1991) *Gaia: The Practical Science of Planetary Medicine*. Gaia Books: London

——, and Margulis, L. (1974) 'Atmospheric homeostasis by and for the biosphere: The Gaia hypothesis.' *Tellus* 26, pp.2–10

Margulis, L. (1999) *The Symbiotic Planet: A New Look at Evolution*. Weidenfeld and Nicolson: London

Markos, A. (2002) *Readers of the Book of Life*. Oxford University Press

Marmot, S.G., Syme, S.L, Kagan, A. (1975) 'Epidemiologic studies of coronary heart disease and stroke in Japanese men living in Japan, Hawaii, and California: prevalence of coronary and hypertensive and associated risk factors.' *American Journal of Epidemiology* 102, pp.514–25

Maturana, H.R., and Varela, F.J. (1987) *The Tree of Knowledge: The Biological Roots of Human Understanding*. Shambhala Publications Inc.: Boston

Mead, G.H. (1934) *Mind, Self and Society.* University of Chicago Press
——, (1936) *Movements of Thought in the Nineteenth Century.* University of Chicago Press
——, (1938) *The Philosophy of the Act.* University of Chicago Press
Merleau-Ponty, M. (1945/1976) *Phénoménologie de la Perception.* Gallimard: Paris
Moerman, D. (2002) *Meaning, Medicine and the Placebo Effect.* Cambridge University Press
Murray, N. and Holman, M. (2001) 'The role of chaotic resonances in the Solar System.' *Nature* 410, pp.773–79
Myin, E. and O'Regan, J.K. (2002) 'Perceptual consciousness, access to modality and skill theories; a way to naturalize phenomenology?' *J. Consciousness Studies* 9, pp.27–45
Naydler, J. (1996) *Goethe on Science: An Anthology of Goethe's Scientific Writings.* Floris Books: Edinburgh
Ornish, D. (1998) *Love and Survival: How Good Relationships Can Bring You Health and Well-Being.* Vermilion: London
Orr, D. (1994) *Earth in Mind: On Education, Environment, and the Human Prospect.* Island Press: Washington DC
——, (2004) *Nature of Design: Ecology, Culture, and Human Intention.* Oxford University Press
O'Toole, P.R., Robinson, S., and Myerscough , M.R. (1999) 'Self-organized criticality in termite architecture: a role for crowding in ensuring ordered nest expansion.' *J. Theoret. Biol.* 198, pp.305–27
Peng, C.-K., Mietus, J., Hausdorf, J.M., Havlin, S., Stanley, H.E., Goldberger, A.L. (1993) 'Long-range anticorrelations and non-Gaussian behavior of heartbeat.' *Phys. Rev. Lett.* 70 pp.1343–46
——, Havlin, S., Stanley, H.E., and Goldberger, A.L. (1995) 'Quantitative analysis of heart rate variability,' pp.82–87 in *Dynamical Disease: Mathematical Analysis of Human Illness,* J.Belair, L. Glass, U. an der Heiden, and J. Milton (eds) Springer Verlag: Los Angeles
——, Havlin, S., Stanley, H.E., and Goldberger, A.L. (1996) 'Fractal mechanisms and heart rate dynamics: Long-range correlations and their breakdown in disease.' *J. Electrocardiol.* 28, pp.59–65
Phillips, D.P., Ruth, T.E., Wagner, L.M. (1993) 'Psychology and survival.' *The Lancet* 342: pp.1142–5
Pivar, S. (2007) *Urform: An Inconvenient Theory.* Dalton Press: NY.
Poon, C.-S., and Merrill, C.K. (1997) 'Decrease of cardiac chaos in congestive heart failure.' *Nature* 389, pp.492–95
Proskauer, G. (1986) *The Rediscovery of Colour.* Anthroposophic Press: Spring Valley, New York

Quist, D. and Chapela, I.C. (2001) 'Transgenic DNA introgressed into tradi-
    tional landraces in Oaxaca, Mexico.' *Nature* 414, pp.541–43
Rasnick, D. (2000) 'Auto-catalyzed progression of aneuploidy explains
    the Hayflick limit of cultured cells, carcinogen-induced tumors in
    mice, and the age distribution of cancer.' *Biochem. J.* 348,
    pp.497–506
——, and Duesberg, P.H. (1999) 'How aneuploidy affects metabolic control
    and causes cancer.' *Biochem. J.* (1999) 340, pp.621–30
Reason, P. (1998) 'Co-operative inquiry as a discipline of professional practice.'
    J. Interprofessional Care 12, pp.419–36
Reason, P. and Goodwin, B.C. (1999) 'Toward a science of qualities in organ-
    izations: Lessons from complexity theory and postmodern biology.'
    Concepts and Transformation 4, pp.281–317
Reibe, N. and Steinle, F. (2002) 'Exploratory Experimentation: Goethe,
    Land and Color Theory.' *Physics Today,* 55, pp.43–50
Richards, Robert (2002) *The Romantic Conception of Life: Science and
    Philosophy in the Age of Goethe.* University of Chicago Press
Romijn, H. (2002) 'Are virtual photons the elementary carriers of conscious-
    ness?' *J. Consciousness Studies* 9, pp.61–81
Seamon, D. and Zajonc, A. (eds) (1998) *Goethe's Way of Science: A
    Phenomenology of Nature.* State University of New York Press
Seeley, T. D. (1995) *The Wisdom of the Hive.* Harvard University Press
Senge, P. (1990) *The Fifth Discipline: The Art and Practice of the Learning
    Organization.* Doubleday: New York
Shaw, P. (2002) *Changing Conversations in Organizations: A Complexity
    Approach to Change.* Routledge: London
Sheldrake, R. (1981) *A New Science of Life: The Hypothesis of Formative
    Causation.* Blond and Briggs: London
——, (1988) *The Presence of the Past: Morphic Resonance and the Habits of
    Nature.* Collins: London
——, (1990) *The Rebirth of Nature: The Greening of Science and God.*
    Century: London
Shiva, V. (1988) *Staying Alive: Women, Ecology and Development.* Zed Books
——, (1993) *Monocultures of the Mind: Biodiversity, Biotechnology and the
    Third World.* Third World Network
——, (2002) *Water Wars; Privatization, Pollution and Profit.* South End Press
——, (2005) *Earth Democracy.* South End Press
Silberstein, M. (1998) 'Emergence and the mind-body problem.' *J.
    Consciousness Studies* 5, pp.464–482

Solé, R.V., Miramontes, O., and Goodwin, B.C. (1993) 'Oscillations and chaos in ant societies.' *J. Theoret. Biol.* 161, pp.343–57

Solé, R.V. and Goodwin, B.C. (2000) *Signs of Life: How Complexity Pervades Biology.* Basic Books: New York

Song, C., Havlin, S., and Makse, H.A. (2005) 'Self-similarity of complex networks.' *Nature* 433, pp.392–95

Spiegel, D. (1993) *Living Beyond Limits; New Hope and Help for Facing Life-Threatening Illness.* Ballantine Books: New York

Stacey, R.D., Griffin, D., and Shaw, P. (2000) *Complexity and Management: Fad or Challenge to Systems Thinking?* Routledge: London and New York

Stewart, I. (1989) *Does God Play Dice?* Penguin: Harmondsworth, UK

Taylor, J.C. (2001) *The Hidden Unity of Nature's Laws.* Cambridge University Press

Thiessen, S., and Saedler, G. (2001) 'Floral Quartets.' *Nature* 409, pp.469–71

Thompson, D'Arcy Wentworth (1917, 1961) *On Growth and Form.* Cambridge University Press

Varela, F.J., Thompson, E., and Rosch, E. (1993) *The Embodied Mind: Cognitive Science and Human Experience.* The MIT Press: Cambridge, Mass.

Webster, G.C. and Goodwin, B.C. (1997) *Form and Transformation.* Cambridge University Press

Wemelsfelder, F., Hunter, E.A., Mendl, M.T., and Lawrence, A.B. (2000) 'The spontaneous qualitative assessment of behavioural expressions in pigs: First exploration of a novel animal welfare measurement.' *Applied Animal Behaviour Science* 67, pp.193–215

Wemelsfelder, F. (2001) 'The inside and outside aspects of consciousness: Complementary approaches to the study of animal emotion.' *Animal Welfare* 10, pp.129–139

Whitehead, A.N. (1925, 1967) *Science and the Modern World.* Free Press: New York

——, (1929, 1967) *Process and Reality: An Essay in Cosmology.* Free Press: New York

Wilkins, A.S. (2002) *The Evolution of Developmental Pathways.* Sinauer Associates

Woodman, M. (1982) *Addiction to Perfection: The Still Unravished Bride.* Inner City Books: Toronto

Zhang, L., Zhou, W., Velculescu, V.E., Karn, S.E., Hruban, R.H., Hamilton, S.R., Vogelstein, B., and Kinzier, K.W. (1997) 'Gene expression in human gut cancer cells.' *Science* 276, pp.1268–72

Index

Floris
Books

For news on all the latest books, and to get exclusive discounts, join our mailing list at:

florisbooks.co.uk/mail/

And get a FREE book
with every online order!

We will never pass your details to anyone else.

Printed in the USA
CPSIA information can be obtained
at www.ICGtesting.com
JSHW011519221024
72172JS00009B/76

9 780863 155596